Issues in International Climate Policy

Theory and Policy

Edited by

Ekko C. van Ierland

Professor of Environmental and Natural Resource Economics, Wageningen University, The Netherlands

Joyeeta Gupta

Associate Professor, Institute for Environmental Studies, Vrije Universiteit Amsterdam, The Netherlands

Marcel T.J. Kok

Researcher, RIVM National Institute for Public Health and Environment, Office for Environmental Assessments, The Netherlands

Edward Elgar
Cheltenham, UK • Northampton, MA, USA

Published by
Edward Elgar Publishing Limited
Glensanda House
Montpellier Parade
Cheltenham
Glos GL50 1UA
UK

Edward Elgar Publishing, Inc.
136 West Street
Suite 202
Northampton
Massachusetts 01060
USA

A catalogue record for this book
is available from the British Library

Library of Congress Cataloguing in Publication Data

Issues in international climate policy: theory and policy/edited by Ekko C. van
Ierland, Joyeeta Gupta, Marcel T.J. Kok.
        p. cm.
    Includes index.
    1. Climatic changes—Government policy. I. Ierland, E.C. van (Ekko C.
van). II. Gupta, Joyeeta, 1964–   . III. Kok, Marcel T.J., 1968–
QC981.8.C5 I875   2003
363.738'7—dc21   2002034675

ISBN 1 84376 191 2

Printed and bound in Great Britain by MPG Books Ltd, Bodmin, Cornwall

Contents

PART III IMPLEMENTATION AND THE DEVELOPMENT OF A
CLIMATE REGIME

Figures

Tables

Boxes

Contributors

Henk Addink, Associate Professor, Faculty of Law, Utrecht University, The Netherlands

Magnus Andersson, Department of Environmental Policy, Wageningen University, The Netherlands

Bas Arts, Lecturer in International Environmental Politics, University of Nijmegen, The Netherlands

Jan Bandsma, Faculty of Economics, University of Groningen, The Netherlands

Marcel M. Berk, Senior Policy Analyst, National Institute for Public Health and Environment (RIVM), Office for Environmental Assessment, The Netherlands

Jos Cozijnsen, Consulting Attorney on issues of energy and environment, Utrecht, The Netherlands

David J.-E. Grimeaud, Research Associate, Transnational Legal Research Institute (METRO), Maastricht University, The Netherlands

Joyeeta Gupta, Associate Professor, Institute for Environmental Studies, Vrije Universiteit, Amsterdam, The Netherlands

Jaap C. Jansen, Senior Scientist, Energy Research Centre, Petten, The Netherlands

Catrinus J. Jepma, Professor, Faculty of Economics, University of Groningen and University of Amsterdam

Marcel T.J. Kok, Researcher, National Institute for Public Health and Environment (RIVM), Office for Environmental Assessment, The Netherlands

Onno Kuik, Institute for Environmental Studies, Vrije Universiteit, Amsterdam

Ton Manders, CPB Netherlands Bureau for Economic Policy Analysis, National Institute for Public Health and Environment (RIVM), The Netherlands

Bert Metz, National Institute for Public Health and Environment (RIVM), Office for Environmental Assessment, The Netherlands

Arthur Mol, Professor, Environmental Policy, Wageningen University, The Netherlands

Gert-Jan Nabuurs, Alterra Instituut, Wageningen, The Netherlands

Andries Nentjes, Professor, Faculty of Law, Groningen University, The Netherlands

Peter Nijkamp, Professor, Departement of Spatial Economics, Vrije Universiteit, Amsterdam, The Netherlands

Richard S.J. Tol, Professor, Centre for Marine and Climate Research, Hamburg University, Germany

Willemijn Tuinstra, National Institute for Public Health and Environment (RIVM), Office for Environmental Assessment, The Netherlands

Wytze van der Gaast, Foundation Joint Implementation Network, The Netherlands

Juliette van der Jagt, PhD candidate, Institute of Constitutional and Administrative Law, Utrecht University, The Netherlands

Ekko C. van Ierland, Professor, Environmental Economics and Natural Resources Group, Wageningen University, The Netherlands

Jelle G. van Minnen, National Institute for Public Health and Environment (RIVM), Office for Environmental Assessment, The Netherlands

Harmen Verbruggen, Professor, Institute for Environmental Studies, Vrije Universiteit, Amsterdam, The Netherlands

Edwin Woerdman, PhD candidate, Faculty of Law, Groningen University, The Netherlands

1. Options for international climate policies: towards an effective regime

Ekko C. van Ierland, Joyeeta Gupta and Marcel T.J. Kok

1 BEYOND BONN AND MARRAKESH

Climate change has evolved from being an abstract scientific problem to take centre stage as one of the most politically, economically and environmentally challenging issues of modern life. In 1992, the United Nations Framework Convention on Climate Change (FCCC) was adopted. It included principles and policies that were to guide the international community in dealing with the climate change issue. In 1997, the Kyoto Protocol to the FCCC was adopted with specific quantitative commitments for the developed countries and mechanisms to promote the implementation of these commitments. However, these commitments brought a price tag with them and made ratification of the Kyoto Protocol a major problem. The US hesitated for a long time, and early in his term of office President G.W. Bush made it clear that his country was going to withdraw from the negotiations. At the same time, the other countries within the FCCC tried at the sixth and seventh Conference of the Parties (CoP-6bis and CoP-7) to solve many of the outstanding issues to enable ratification of the Kyoto Protocol before the Rio+10 Earth Summit in Johannesburg in 2002. There is now more clarity about how the quantitative commitments are to be implemented, and countries are ratifying the Kyoto Protocol in quick succession.

It is clear that the existence of a common enemy, the climate change problem, is not in itself enough to unite the world in a common commitment to deal with that problem. Realists have long asserted that states operate in the international arena to defend their own narrowly-defined self interests. But realist theories could not explain the continuing cooperation in the international arena when, despite the benefits of free-riding, countries were cooperating with each other to develop frameworks and institutions for dealing with complex problems. Short-term economic interests alone could not explain the growing support for international institutions and instruments.

1

This led regime analysts to conclude that countries were getting socialized into the international process and cooperation was becoming institutionalized. Membership of international society brings with it certain responsibilities and certain opportunities. A stable international legal order brings many benefits to countries, especially as life in one country becomes increasingly interlinked with life elsewhere through the complex process of globalization.

As regards climate change, we now have agreement about the rules of the game in relation to the Kyoto Protocol. But the issue is becoming more and more complex; so much so that the decisions taken at Marrakesh in 2001 ran to 150 pages and it takes considerable scientific understanding to see what the implications are for individual countries. Besides that, the long-term dimensions of the climate problem may remain on the political agendas. Deeper cuts in emissions than those envisaged in the Kyoto Protocol will be needed to stabilize the concentrations of greenhouse gases. The critical question is: how stable is the climate regime? Have the existing agreements set the stage for a legitimate and effective long-term process which is able to realize the global emission reductions deemed necessary to achieve safe concentration levels of greenhouse gases in the atmosphere?

The complexity of the climate problem itself and of international relations and the negotiation process means that there are a large number of outstanding issues in international climate policy. Despite the pace of the negotiations, some of these issues have remained on the agenda for a remarkably long time.

This book explores some of the outstanding issues on the basis of research that has been conducted in the Netherlands over the last few years. The book takes a longer-term perspective, one that goes beyond the first commitment period of the Kyoto Protocol. The research results presented in this book reflect the high degree of interest in the international dimensions of climate change policies in the Dutch research community. A considerable part of this research has been conducted within the framework of the Dutch National Research Programme on Global Air Pollution and Climate Change (NRP). This programme also took the initiative to bring together a variety of research disciplines and perspectives concerned with these issues and to synthesize the findings of the research in this volume. The book contributes to the discussion on international climate policies by providing an in-depth analysis of the main characteristics of the problem of climate change (Part I), various potential solutions to the problem and their consequences (Part II), and the development and implementation of the international climate regime (Part III). In addition to providing an analysis of the main characteristics of the climate change issue and the relevant international context, a central aim of the book is to present the various policy strategies in a long-term perspective.

The book sheds light on the positions of the various actors and their specific interests. A careful analysis of the various policy options and their implications is presented to give a clear overview of the choices that have to be made in the near future. Finally, the book analyses the challenges for the actual implementation of an international policy regime in practice. Specific attention is given to a study that aimed to analyse the implications of very strict policies by means of interactive dialogue with stakeholders from various economic sectors.

2   BACKGROUND: AN HISTORICAL OVERVIEW

During the last decade, the international community has negotiated the United Nations Framework Convention on Climate Change in 1992 (FCCC, 1992) and the Kyoto Protocol in 1997 (KPFCCC, 1997). The aim of the Framework Convention is to stabilize the concentration of greenhouse gases in the atmosphere at a safe level (Article 2). It identifies principles that should govern the division of responsibility and the implementation process (Article 3) and policies and measures (Article 4) that need to be taken by developed (Annex I) and developing countries to deal with the issue of climate change. It also establishes a financial mechanism, which was to be operated on an interim basis by the Global Environmental Facility (Articles 11 and 21). Following the entry into force of the Convention, negotiations began on the Kyoto Protocol which was adopted in 1997. This Protocol goes a step further than the FCCC, in that it articulates legally binding emission-related commitments for the developed countries (Article 3) and identifies three mechanisms for facilitating the implementation of the Protocol. These are joint implementation (JI) (Article 6), the Clean Development Mechanism (CDM) (Article 12) and emission trading (Article 17). It also lists a number of general obligations for both developed and developing countries.

From a legal and political perspective, the Protocol is a considerable step forward (Yamin, 1998). From an economic perspective, the Protocol apparently has serious implications since the US Senate originally linked ratification of the Protocol to 'meaningful participation' by developing countries (Clinton, 1997). The European Union (EU) countries and other countries in the developed world were initially unwilling to unilaterally commit themselves to the process through ratification. However, in 2000, the EU decided to push the issue of ratification so that the Kyoto Protocol could enter into force by 2002. During CoP-6 in The Hague (2000) governments failed to reach agreement on the implementation of the provisions included in the Kyoto Protocol. The newly elected president of the United States, G.W.

Bush, then declared in early 2001 that the US would withdraw from the Kyoto Protocol negotiations. The resumed negotiations at CoP-6bis took place in a dramatically changed context where the rest of the world was determined to show that it could make progress on the climate-change issue even without US participation.

From an environmental perspective, however, the Protocol is far from satisfactory. The target for emission reduction of 5.2 per cent relative to the emission levels of 1990 for the developed countries jointly is seen as less than what is needed as a first step to deal with the problem of climate change, and falls short of initial assessments that, in the longer term, the emission levels of CO_2, for instance, need to fall by more than 60 per cent to stabilize global concentrations at 1990 levels (Houghton et al., 1990, Chapter XVIII). The inclusion of six gases and sinks in the Kyoto Protocol weakens both the target and the environmental impact of the Protocol considerably. Furthermore, the inclusion of the Clean Development Mechanism is expected, in effect, to 'inflate' the target of the developed countries.[1] The current economic situation in Eastern Europe, especially Russia and Ukraine, combined with their allegedly high emission allowances ('assigned amounts') raises the issue of trading 'hot air' (Grubb and Hourcade, 2000). All in all, there are a large number of loopholes in the Protocol. The apparent lack of policy commitment to the process, in particular on the part of the oil producers and many of the other developing countries, is also a major impediment. The political deal at CoP-6bis was criticized because it further reduced the Kyoto Protocol's environmental integrity, although others praised the fact that key parties (such as the EU) now showed considerable flexibility in their positions to make an agreement possible.

From the perspective of the developed countries, attempts to reduce emissions within their own countries are seen as very expensive. This explains the search for cost-effective mechanisms that will encourage the private sector to invest domestically and abroad in emission-reduction opportunities. IPCC Working Group III (2001c), however, concluded that in the absence of emission trading between Annex I countries, the majority of global studies show reductions in projected GDP of about 0.2 to 2 per cent by the year 2010; with full emission trading the reductions in GDP are between 0.1 and 1.1 per cent of projected GDP. Furthermore, these countries fear that the emission level of the developing countries will rapidly negate any emission reduction undertaken in the North. Hence the pressure to at least keep the key developing countries on board with concrete quantitative commitments.

From the perspective of the developing countries, the climate change problem has primarily been caused by the developed countries, which are also seen to have the capability to reduce their own emissions. Mexico and South Korea are no longer members of the Group of 77, so these countries operate outside the ambit of that Group. The key members of OPEC are promoting their own national interests and opposing, to the extent that they dare,[2] concrete commitments for the North and compliance mechanisms. For the 42 small island states and 53 African countries, the key concern is: how do we cope with the potential impacts of climate change? For the remaining developing countries, the concerns are more ambiguous, in that many of them have an active industrialized sector whose economic interests the governments wish to protect, while at the same time they also have a considerable agricultural and tourism sector that may be seriously affected by the potential impacts of climate change. Most of the developing countries do not feel the climate-change regime substantively reflects the principles enshrined in the Convention.

If we move from the debate on the outcomes of the negotiations to the issue of actual emission levels, there are indications that the developed countries had indeed brought their collective emissions back to around 1990 levels by the year 2000. Bert Bolin (1998), chairman emeritus of IPCC, showed that there were reductions in the emissions of some countries and increases in others, with an average reduction of 5 per cent in that period. However, credit for this goes primarily to the collapse of the economies of the East Bloc countries, the privatization of the coal-mining sector in the United Kingdom and the unification of Germany. Most of the other countries have experienced considerable emission increases in the period (see for example European Commission, 1998a). Furthermore, although most of these countries are implementing their commitments to submit national communications to the Secretariat, they are late and incomplete (FCCC/SBI/1998/INF.3/Add.1).

The Global Environmental Facility (GEF) has provided assistance to developing countries for the purpose of capacity-building in relation to climate change. There are 186 Activities Implemented Jointly Projects, although the majority of these projects have been in the East Bloc countries (JIQ, 2002) despite the fact that such projects were meant to be focused on all parts of the world. Thirty developing countries have submitted their national communications to the Secretariat. Although many initiatives have been taken in the area of technology transfer, it is argued by many parties involved that actual implementation of this obligation is not visible.

3   THE OUTSTANDING KEY ISSUES

Against this background, the key issues that are outstanding in the context of the climate change regime are (i) the articulation of the precautionary principle in relation to emission reductions in future commitment periods; (ii) the link between climate change and other international regimes; (iii) the economic aspects of the mechanisms in the Protocol; (iv) the issue of equity and differentiation between countries; (v) a closer examination of the 'sink' issue; (vi) compliance and (vii) implementation-related issues and the role of non-governmental actors in influencing the negotiations. These topics are the main focus of the book.

This book is in three parts. Part I (Problem Exploration and Relevant Context) analyses various dimensions of the climate problem as an environmental as well as a societal and policy problem. Chapter 2 examines the need for climate policy. Chapter 3 looks at scenario analyses and identifies the problem that lies ahead in the coming decades. To show that climate change is only one problem in a rapidly changing international arena, Chapter 4 is devoted to identifying global trends and the specific problems that are related to climate change, both to its causes and its potential solutions. Chapter 5 deals with the question of how transboundary pollution and climate change can be dealt with in a rapidly changing international policy context of globalization, increased international trade and harmonization of policies.

Part II (Towards Solutions and Consequences of these Solutions) focuses on the various policy measures that may contribute to mitigation and adaptation. Chapter 6 deals with the questions of policies and measures. Chapter 7 discusses the potential of and problems with flexible instruments, which play such a dominant role in the Kyoto Protocol. Chapter 8 considers options for using carbon sinks to reduce the net greenhouse gas emissions. Specific attention is paid to the background and the fundamental aspects of including sinks in the international policy options and the implications for JI and CDM. The chapter focuses on the role of carbon sinks in international climate policy, as compared to more traditional policy options like energy conservation and fuel switch.

Part III (Implementation and the Development of a Climate Regime) deals with some of the complications of implementing policies in practice. Not surprisingly, the issue of burden differentiation and burden sharing are at the heart of the international negotiations. Chapter 9 focuses on the distribution of costs and benefits of climate change and on the various ways of developing a comprehensive and fair climate regime beyond the ad hoc approach of the Kyoto Protocol. Chapter 10 deals with the impacts of climate change policies on international trade and the specific impacts on developing countries. Chapter 11 focuses on the legal aspects of the Kyoto Protocol and the options for enforcement of the rules in the event of non-compliance. Chapter 12 discusses the role of NGOs in the climate debate, both as actors influencing public opinion and as actors influencing the international negotiations and the parties to the conference. Chapter 13 is devoted to identifying possible strategies for far-reaching emission reductions in an interactive discourse with various stakeholders in the climate change debate, based on an extensive study in the Netherlands.

4   AN EVALUATION OF THE KEY ISSUES

Climate change is a serious issue and its impacts are likely to be serious. The recent scientific assessment of the Inter-governmental Panel on Climate Change (IPCC, 2001a, b) refers to the unprecedented nature of current emission levels and the possible impacts on unique and vulnerable systems, and the likelihood of major discontinuities in the climatic system. A number of scenarios have been developed on the basis of story lines for the future by the IPCC. If one compares the monetary damage assessments of the possible impacts and the calculated costs of emission reduction on the basis of standard neo-classical cost–benefit analysis, one might be tempted to argue that the potential benefits do not justify the costs of taking action, as Manders and Tol point out (see Chapter 3). On the other hand, if one looks at the issue of climate change in wider economic terms, in terms of the potential consequences for the income gap between rich and poor countries, in terms of the polluter pays principle and in terms of the need to protect and conserve unique and vulnerable systems, it is clear that a narrowly defined cost–benefit analysis is far from adequate as a basis for designing policies to reduce greenhouse gas emissions and increase sinks. The international political system is so closely integrated that a widening economic gap between the rich and the poor may have longer-term consequences for global prosperity, economic growth and health risks, and may lead to civil crises, failures in governance of countries and growth in the numbers of environmental, political and economic refugees. From such a long-term perspective, it is imperative to take into account the full costs to society of inaction in the field of climate change. Evidently, there are plenty of domestic and international reasons that justify policies and strategies to encourage an industrial transformation towards a low-emissions economy; but most countries face serious dilemmas (see Chapter 2).

At the same time, it is appropriate to reflect on whether the current globalization trends work against or support climate change policies. Nijkamp and Verbruggen (see Chapter 4) identify the megatrends in the global village as institutional change, internationalization and economic integration, rapid technological progress, the emergence of a knowledge economy, improved logistic and transport systems, demographic developments and transformation, cultural shifts and new international business strategies. These megatrends are likely to export western energy-intensive production patterns, modes of consumption, transportation modes and lifestyles to the rest of the world, thus rapidly increasing the greenhouse-gas emissions of other countries in the world. At the same time, others argue that the globalization trends may lead to a widening gap between the rich and the poor, with the growth of selective labour migration. Capital flight to centres of economic power and monopoly positions held by companies based mainly in the North may damage local cultural infrastructure and institutions and lead to the closure of small businesses and industries. It may marginalize the small farmer and may lead to further disempowerment of the poor (Amin, 1993; Madeley, 2000; De Rivero, 2001; Petrella, 2000). The combination of a growing global middle class with conspicuous consumption patterns on the one hand, and the further marginalization and exclusion of many of the world's poor on the other, may lead to rapid increases of emissions of greenhouse gases and to a reduced ability to cope with the potential impacts of climate change.

At the same time, institutional change is gradually taking place. The problem of climate change has reinforced the notion that we live in one world. It is leading to an internationalization of norms and values, technological and environmental performance standards and to the use of environmental resources as a major bargaining chip between North and South. More importantly, as Addink et al. point out (see Chapter 5), the neo-liberalization processes may be expanding more rapidly than the evolution of global institutions to regulate international life. There is a danger that the international policymaking process will lag behind the economic mechanisms that are a driving force behind the global economy, making it constantly harder for the policy community to cope with the economic system. Although we are moving away from the nation-state model in which society was seen as manageable, we still have a long way to go towards a system of effective multi-level and multi-actor governance. We observe a number of legal innovations in international law which allow for the combination of binding norms with flexible instruments, and soft managerial approaches with strong enforcement approaches. It will, however, be a long and complex process to find a balance in the governing process.

Coming to the specifics of the climate change regime, a critical question has always been whether policies to deal with climate change should be price- or quantity-based. While price-based systems appear to be much more politically attractive (taxes often seem politically unpalatable) the complex nature of the design of the Kyoto Protocol mechanisms leads Jepma and Bandsma (see Chapter 6) to conclude that price-based systems will be much easier to implement in the long run and be more effective. Whether this analysis is academic in nature or not remains to be seen, given the extent to which the quantity-based systems have advanced in the Kyoto Protocol.

Woerdman et al. (see Chapter 7) go into the specifics of the different mechanisms and analyse the key potential problems and political barriers. They conclude that the quantity-based systems in the Kyoto Protocol do indeed reduce the price of compliance and implementation considerably; but they also conclude that trading in hot air with Eastern and Central European countries does not lead to genuine emission reductions. They argue that in the second commitment period further leakage can be reduced if developing countries are willing to accept quantitative commitments.

The role of sinks in reducing the concentrations of greenhouse gases in the atmosphere is a challenging issue. That the protection and extension of sinks contributes to mitigation of the climate change problem is undisputed. But the way in which sinks were introduced in the Kyoto Protocol raises questions about the environmental effectiveness of the climate change regime. There are a number of political complications, such as the permanence of sinks, their impact on the incentives to reduce emissions and their impact on countries that have large land resources and sinks. Chapter 8 by van Minnen, van Ierland and Nabuurs concludes that between 1 and 2 Gt of carbon can be stored per year in sinks, depending on the definition used.

Within the present context of the Kyoto Protocol with its modest targets, flexible mechanisms and the potential to use sinks to meet national commitments, the question is how should the regime develop further? It is clear that for any long-term regime it is vital that all countries are ready to commit in one way or another to measurable targets. Berk et al. (see Chapter 9) argue that in order to secure commitment to a well-elaborated climate regime it is vital that the regime is perceived as legitimate, predictable and fair. This calls for the development of a comprehensive approach based on comparable principles for all parties. While fairness is seen as a debatable issue given the multiple definitions available on the issue and that different countries employ different definitions on the subject of burdensharing, the need for predictability in itself calls for a comprehensive approach. The chapter evaluates a number of approaches and concludes that the time is ripe for international negotiations on how the 'common but differentiated approach' is to be articulated in the regime. The chapter concludes that the precedents in the regime thus far are not coherent or consistent.

Although the climate change regime is being developed as if it operates in isolation, it clearly comes in addition to existing international agreements in other fields. One critical regime with which there are close connections is the international trade regime. Kuik et al. (see Chapter 10) argue that in the trade regime countries are not allowed to discriminate between foreign suppliers or between foreign and domestic suppliers of a similar product. This reduces the scope for countries to develop national policies to promote sustainable production processes and at the same time prevents countries from using climate policy for strategic purposes, such as protecting domestic production.

The future success of the regime depends on how well the Kyoto Protocol is implemented during the first commitment period. The high stakes involved may lead countries to default on their obligations. The adoption of quantitative commitments in combination with the flexible mechanisms calls for a well-developed compliance regime that goes beyond ensuring that countries have adopted policies to implement their commitment, but that they have indeed complied with the requirements of the regime. This applies especially because the incentive to free-ride is minimized when there is clarity about what the consequences of non-compliance will be. Van der Jagt (see Chapter 11) argues that the Kyoto Protocol has all the necessary elements for establishing a comprehensive compliance regime. In the follow-up to the Kyoto Protocol, a Compliance Committee, a facilitation branch and an enforcement branch have been established. Responses to non-compliance have been adopted and rules of procedure drawn up. Whether this compliance regime is legally binding, however, remains an open question.

How do social actors influence the process of climate negotiations? Arts and Cozijnsen (see Chapter 12) show that the increasing involvement of the business sector and of green non-governmental organizations in the 1990s has provided a strong impetus to the development of the regime. While both groups had diametrically opposing positions and strategies in the early days of the regime, the chapter optimistically argues that, over time, coalitions between environmental groups and industry have formed, as industry increasingly recognizes the need for action and as environmental groups realize that industry has to be part of the solution. The authors warn, however, that environmental groups should not become so co-opted into the process that they are no longer functioning as watchdogs in the regime!

A lot of progress has been made in the design and implementation of an international climate regime, but there is a long way to go. Short-term political strategies stand in the way of comprehensive, long-term cooperative approaches. One recent project (Metz et al., Chapter 13, below) evaluated the possibility of an 80 per cent reduction of emissions by 2050 and used backcasting and participative integrated assessment as a way to explore future emission reductions. The aim of this project was, in a dialogue between stakeholders, policy makers and scientists, to develop visions on what long-term strategies for drastic emission reductions might look like.

5   CONCLUSION

The key challenge for the climate policymakers remains that of developing a legitimate and effective climate regime, which will be able to deal with the long-term challenges the climate problem is posing. With the agreement in Marrakesh an important milestone in the international negotiations has been reached. After the adoption of the FCCC in 1992 the world witnessed a period of divergence and 'artificial' convergence in the international negotiations, resulting in the Kyoto Protocol in 1997. But the Kyoto Protocol does not imply that future negotiations will be easy. The rules of the game were unsettled in Kyoto, and as a consequence the post-Kyoto negotiations diverged again. Between Kyoto and Marrakesh the Kyoto commitments have been re-negotiated, while in the process apparently technical, but in reality politically contentious, issues have been settled. In the coming years the negotiations will tend to diverge again when future commitments have to be negotiated. The political arena of negotiation needs to simplify the complex nature of the subject to the extent that it becomes comprehensible for negotiators. Negotiating very complex agreements, the details of which the bulk of the negotiators may possibly not be able to understand, does not increase the legitimacy of the regime.

Although it is impossible to cover all aspects of climate change, this book presents an overview of the most important issues related to international climate change policies. After many years of scientific research and international negotiations we now have a greater insight into the most important implications of the risk of climate change and the various policy options, both for climate mitigation and for adaptation. We hope this volume will contribute to a better understanding of the issues at stake and in the long run to the development and implementation of a well-designed international policy regime for climate change.

NOTES

1. For example the US State Department issued a Press Release to the effect that the 6 per cent commitment of the US was, in effect, a much weaker commitment.
2. These countries do not wish to be isolated within the G77 and have known, since the first Conference of the Parties, that they will be if they are not careful in their strategy (Mwandosya, 1999)

REFERENCES

Africa–Europe Summit (2000), *Cairo Plan of Action*, Cairo, OAU and EU, 3–4 April 2000.

Amin, S. (1993), 'The Challenge of Globalisation', in South Centre (ed.), *Facing the Challenge: Responses to the Report of the South Commission*, London: Zed Books, pp. 132–8.

Bolin, B. (1998), 'The Kyoto Negotiations on Climate Change: A Science Perspective', *Science*, **279**, 330–51.

Clinton, W.J. (1997), Remarks by the President on Global Climate Change at the National Geographic Society on October 22, 1997; also available at http://www.whitehouse.gov/ Initiatives/Climate/19971022-6127.html.

De Rivero, O. (2001), *The Myth of Development*, London and New York: Zed Books.

European Commission (1998a), *Second Communication from the European Community Under the UN Framework Convention on Climate Change*, Brussels: Commission of the European Communities (SEC(98) 1770).

European Commission (1998b), *Climate Change – Towards an EU Post-Kyoto Strategy, Communication from the Commission to the Council and the European Parliament*, Brussels: Commission of the European Communities (COM (98) 353).

European Environment Council (2000), *Council Conclusion on the Community Strategy on Climate Change*, June 2000, Brussels: Commission of the European Communities.

Franck, T.M. (1995), *Fairness in International Law and Institutions*, Oxford: Oxford University Press.

Grubb, M. and J.C. Hourcade (2000), 'Implementing EU Commitments Under Kyoto', in J. Gupta and M. Grubb (eds), *Climate Change and European Leadership: A Sustainable Role for Europe?*, Dordrecht: Kluwer Academic Publishers, pp. 239–60.

Houghton, J.T., G.J. Jenkins and J.J. Ephraums (1990), *Climate Change: The IPCC Scientific Assessment*, Cambridge: Cambridge University Press.

Intergovernmental Panel on Climate Change (IPCC) (2001a), J.T. Houghton, Y. Ding, D.J. Griggs, M. Noguer, P.J. van der Linden, X. Dai, K. Maskell and C.A. Johnson (eds), *Climate Change 2001. The Scientific Basis*, IPCC Working Group I, UK/USA: Cambridge University Press.

Intergovernmental Panel on Climate Change (IPCC) (2001b), J. McCarthy, O.F. Canziani, N.A. Leary, D.J. Dokken and K.S. White (eds), *Climate Change 2001. Impacts, Adaptation, and Vulnerability*, IPCC Working Group II, UK/USA: Cambridge University Press.

Intergovernmental Panel on Climate Change (IPCC) (2001c), B. Metz, O. Davidson, R. Swart and J. Pan (eds), *Climate Change 2001. Mitigation*, IPCC Working Group III, UK/USA: Cambridge University Press.

JiQ (2002), Planned and ongerty AIJ Pilot Projects, Joint Implementation Quarterly, Volume 8, No.1, April 2002, p.14.

Japan–EU Summit (2000), *Joint Conclusions*, EU Press Release 10577/00, Brussels: Commission of the European Communities.

Kyoto Protocol to the United Nations FCCC, December 1997, FCCC/CP/ 1997/L.7/Add.1

Madeley, J. (2000), *Hungry for Trade: How the Poor Pay for Free Trade*, London: Zed Books.

Mwandosya, M.J. (1999), *Survival Emissions: A Perspective from the South on Global Climate Change Negotiation*, Tanzania: DUP (1996) LIMITED and The Centre for Energy, Environment, Science and Technology (CEEST-2000).

Oberthür, S. and H. Ott (1999), *The Kyoto Protocol. International Climate Policy for the 21st Century*, Berlin: Heidelberg et al., Springer Verlag, p. 359.

Petrella, R. (2000), *The Water Manifest: Arguments for a World Water Contract*, London: Zed Books.

United Nations Framework Convention on Climate Change, (New York) 9 May, 1992, in force 24 March, 1994, 31 I.L.M. 1992, 822.

US State Department (1998), *US State Department Fact Sheet*, Washington: US State Department, January 15, 1998.

Yamin, F. (1998), 'The Kyoto Protocol: Origins, Assessment and Future Challenges', *Review of European Community and International Environmental Law*, 7–2, 113–27.

Young, O. (2001), 'The Behavioral Effects of Environmental Regimes: Collective Action vs. Social Practice Models', *International Environmental Agreements: Politics, Law and Economics*, 1(1), 9–29.

PART I

Problem Exploration and Relevant Context

2.   Why reduce greenhouse gas emissions? Reasons, issue-linkages and dilemmas

Joyeeta Gupta and Richard S.J. Tol[1]

1   INTRODUCTION

Never before have we faced such a complex, uncertain and potentially dangerous global issue as climate change. Never have so many nations rallied together to deal with a problem. And yet, although there has been almost universal participation[2] in the lengthy negotiations,[3] within the complex institutional structure of the climate change negotiations the participants (see Chapter 1) are not necessarily concerned with how to deal with the problem of climate change as how to cope with the process of negotiation without losing too much face. Hence, we see developing countries shying away from a real willingness to state what sort of voluntary measures they are taking and are willing to take (see G-77 position at Kyoto and in post-Kyoto negotiations) and accepting loopholes in negotiation texts that may cause the regime to lose its environmental integrity. We also see developed countries avoiding full national accounting of their national emissions (in relation to bunker fuels, emissions by off-shore islands in other parts of the world: see FCCC/COP/1998/Inf.2), indulging in creative accounting, introducing flexibility mechanisms and new gases (CFC, HFC, and SF6) into the negotiation process and trying to see to what extent the inclusion of national sinks could reduce their emission reduction obligations (see van Minnen et al., Chapter 8, below). By the time of the sixth round of negotiations (COP 6), the European Union, which aspired to leadership in the climate change issue, had deferred ratification of the Kyoto Protocol until there was enough support from other developed countries (Oberthür and Ott, 1999; Gupta and Grubb, 2000; Grubb and Yamin 2001). In 2000 the EU initiated concerted efforts to convince other developed countries that steps were needed to ensure that the Kyoto Protocol enters into force in 2002; but the deadlock in the negotiations in November 2000 slowed the momentum. Early in 2001 the US withdrew support for the Kyoto Protocol. Despite, or perhaps because of this, at the second part of the sixth Conference of the Parties, all other parties adopted a set of political decisions which were further refined at COP 7 in Marrakesh. However, many countries may still have difficulties, now or in the future, in actually ratifying the Protocol and/or adopting far-reaching second percel commitments, particularly without US participation.

The reluctance to take on obligations justifies a re-examination of the reasons why the global community entered into the negotiation process in the first place. Why did it set up a global process if it was not entirely serious about actually reducing emissions? What has changed since the process was first set in motion? This chapter first examines the dimensions of the climate change problem, and then argues why there is a need for action. It addresses the following questions: how serious is the climate change problem, and are there serious climate impacts that justify precautionary action? Are there issue-linkages that could make the problem manageable? What dilemmas do countries face? Are there other factors that could render the obstacles merely temporary hitches to the problem-solving process? Hence, this chapter first discusses the potential impact of climate change from a range of perspectives (Section 2.2), before discussing the potential benefits of taking action in the area of climate change on other issues (Section 2.3). It then examines the dilemmas faced by various groups and countries (Section 2.4), and finally draws some conclusions (Section 2.5).

2   CLIMATE CHANGE AND ITS IMPACTS

The evolution of climate change science does not indicate that the predictions made by scientists a decade ago were not justified. On the contrary, scientists have increasing evidence that the greenhouse gas emissions continue to rise and that the accumulated concentrations of these gases are leading to a change in the equilibrium of the earth's climatic system. The state-of-the-art knowledge on climate change is presented in the five-yearly reports of the Intergovernmental Panel on Climate Change (IPCC). The IPCC (Houghton et al., 1996, p. 4) concludes, on the basis of existing science, that 'the balance of evidence suggests a discernible human influence on the global climate'. Although there have been criticisms of the IPCC results (for example Böttcher and Emsley (eds) 1996), there is growing scientific confidence in the data on emission levels. However, there remain uncertainties about the role of sinks (see van Minnen et al., Chapter 8, below) and the regional spread of impacts. But, given the inherently complicated scientific nature of the discussions, van de Sluijs (1997) suggests that instead of looking for the truth, we may need to be satisfied with transparency and multiple assessments of the risk. The 2001 Third Assessment Report (TAR) of IPCC (IPCC-I, 2001) presents a new assessment of the literature on climate change, its impacts and counter-measures. It concludes that there is adequate evidence of climate change and reason to be concerned about its impact. The Report of Working Group I states that in the 20th century the average global surface temperature has increased by $0.6 \pm 0.2°C$ and the average sea level has risen by 0.1 to 0.2 metres. Snow cover has decreased by 10 per cent since the 1960s and precipitation regimes have changed. Droughts have increased in Asia and Africa. Indeed, there is a growing volume of empirical evidence of the implications of the enhanced greenhouse effect, not only in the weather but also in ecosystems. Can the changes in the climate be attributed to human activities? The IPCC report states that the increase of CO_2 by 31 per cent and of methane by 151 per cent since 1750 reflect rates of growth that have not been experienced for 420,000 years at least. The climate change models show that in the period 1990 to 2100 the average global surface temperature may increase by 1.4 to 5.8°C, rainfall may increase in the mid-to-high northern latitudes, and the sea level may rise by 0.09 to 0.88 metres. The potential impacts of climate change are manifold, widespread and diverse. People evaluate climate change impacts in various ways, reflecting different ways of looking at the problem or different value systems. The Third Assessment Report of IPCC (Smith et al., 2001) group the different viewpoints or reasons for concern in four major classes, to which we add a fifth:

1. Unique and valuable systems
2. Distribution and justice
3. Aggregate impacts
4. Large-scale discontinuities
5. Responsibility

2.1 Unique and Valuable Systems

Climate change is likely to have an impact on all systems, but some are more vulnerable than others. This includes ecosystems, species and people with limited possibilities to adapt, either because they are restricted geographically with little options to migrate (for example island and mountain ecosystems) or because they are highly specialized in specific eco-climatic niches. Climate change will put additional stress on such unique and vulnerable systems, forcing them to adapt to changing circumstances. For a number of systems, the additional stress may lead to further deterioration and marginalization of the system. (Smith et al., 2001). They are also rare and special, and therefore, in human eyes, valuable. To some people, the threat that climate change poses to such systems is reason enough to want to avoid climate change. This viewpoint reflects a strong form of sustainability, large intrinsic and existence values of species and a strong conservationist attitude.

The literature contains many examples of unique and valuable systems that may be threatened by climate change. Examples include glaciers, island and mountain cultures and ecosystems, coastal wetlands and coral reefs (Smith et al., 2001). There is some evidence that the 0.7°C global warming of the past century has already damaged certain systems, including glaciers and coral reefs. Species like butterflies may also be endangered. Any further warming would increase the damage to those systems and increase the number of damaged systems (Smith et al., 2001). Any attempt to assess the damage to these systems in economic terms calls for assumptions about their value and is highly controversial.

There is also a political dimension to the unique systems. There are several small independent island states that are likely to disappear in the event of a severe rise in the sea level. What will happen to these nations? Will other countries be willing to provide these people with land to set up their own nations, or will the people be forced to become immigrants and spread out as sometimes unwanted guests in different countries throughout the world? It is not for nothing that Tuvalu, a state of 24 square kilometres spread over nine islands, has become a recent member of the United Nations, fearing for its very existence.

If climate change is defined as a problem affecting unique environmental and political systems, then the issue of uncertainty in the problem is less relevant, and reducing emissions globally becomes of paramount importance. This perspective is usually espoused by environmentalists and, for example, by the Alliance of Small Island States.

2.2   Distribution and Justice

Climate change does not affect everybody equally. Some will lose a little, some will lose a lot, some may benefit. In general, those whose lives are most challenged – because of poverty, illness, lack of power – have the most to fear from climate change (Smith et al., 2001). These people also emit proportionally less greenhouse gases.

The literature is seldom explicit about the distributional consequences of climate change, but there is a lot of support for the conclusion that impacts are unevenly distributed, and that the worst off suffer most. This case is often made for regions or countries; by implication, the same should hold for distribution within countries (Smith et al., 2001).

Impact studies suggest that with moderate warming some would benefit while others would suffer. With higher warming, the losses would increase while the gains would decline, and eventually turn negative (Smith et al., 2001). It is unclear whether the income gap would widen or narrow. Since the poor are more vulnerable than the rich, climate change can expected to increase income and welfare disparities.

Therefore, climate change can be defined as a problem of distribution and justice. The expectation is that climate change may increase the income gap between the rich and the poor, and that certain groups in society impose risks on others. This is sufficient reason to want to avoid greenhouse gas emissions. This forms the basis of the negotiating position generally held by the G-77 which sees the international order as an economically and politically inequitable order in which the developed governments use the international system to protect their own interests at the cost of the developing countries. Implicitly, greenhouse gas emission reduction by rich countries is a transfer of wealth to poor countries.

Equity also has an intertemporal dimension. Climate change is imposed upon any generation by its predecessors. Climate change may cause harm, and it is a well-accepted moral and legal principle that one should try to avoid doing harm to others. This clear principle is obscured by the fact that emission abatement may also cause harm to future generations. The fact that future generations are not actual beings, but potential ones, whose existence is furthermore contingent on decisions made by the current generation further complicates the ethics. Nonetheless, the widely-accepted sustainability goal says that the current generation should not endanger the abilities of future generations to meet their own needs (see Berk et al. Chapter 9 for a more detailed discussion).

2.3 Aggregate Impacts

The potential impacts of climate change are diverse. Some impacts are positive, others are negative. Some impacts are small, others are large. Some impacts are on important sectors, others are on less important sectors. Impacts occur in different places, at different times. To gain a comprehensive picture of vulnerability to climate change, we have to make the impacts comparable to each other, and aggregate them to a small set of interpretable indicators. Money is the most common indicator.

The literature contains only a few estimates of aggregate impacts of climate change. These estimates indicate large uncertainties, but also a few seemingly robust conclusions. At a few degrees of warming, rich countries may enjoy net benefits, but not much more than a percentage of GDP, while poor countries are likely to have net losses, smallish for most and less than 10 per cent of GDP for most. The world as a whole is perhaps better off, depending on how one aggregates. At higher warming, losses grow bigger and benefits turn to losses. The world as a whole faces net losses (Smith et al., 2001).

If one aggregates impacts, using money as the indicator, one can interpret climate change as an efficiency problem. One can then ask whether money is better spent on greenhouse gas emission reduction, on adaptation to and compensation of climate change impacts or on improving primary health care. The relevant indicator is the marginal net present cost of emission. Estimates of the marginal damage of carbon dioxide vary between $0/tC and $100/tC, but the more acceptable estimates range between $5/tC and $25/tC. As Manders and Tol (Chapter 3, below) show, the costs of emission reduction are not justified by the gains for the globe as a whole; this means that from a narrow cost–benefit approach, far-reaching policies do not seem justified. This viewpoint reflects a belief in human rationality and markets. It is held mainly by neo-classical economists, rationalists, and people to the right of the political spectrum.

2.4   Large-scale Discontinuities

Climate change affects the major regulatory systems of the earth. In the distant past, large and rapid changes occurred. Climate change may induce similar large-scale discontinuities, such as a shut down of the thermohaline circulation or a disintegration of the West-Antarctic Ice-Sheet (Smith et al., 2001).

The literature contains a number of qualitative descriptions of potential large-scale discontinuities, but only a few quantitative results which are largely restricted to physical systems. These quantitative studies confirm the real possibility of large-scale discontinuities, but do not estimate the chances of them occurring or the likely impacts (Smith et al., 2001).

The few available studies suggest, however, that the probability of large-scale discontinuities occurring is rather small for modest warming, but increase in the event of more serious global warming (Smith et al., 2001).

To some people, the threat of major and rapid changes, with unpredictable and irreversible consequences, is reason enough to want to avoid climate change. This viewpoint reflects the precautionary principle and strong risk aversion, but also a conservationist attitude.

2.5   The Responsibility Approach

Climate change is partially caused by the emissions of greenhouse gases by people, companies and countries. Some emit far more than others. The impacts, although uncertain, may be spread throughout the world. Under international law states are obliged to ensure that they do not cause harm to other states. In the landmark 1941 Trail Smelter arbitration, where air pollution from Canada was causing damage in the US, it was decided that: 'Under the principles of international law no state has the right to use or permit the use of territory in such a manner as to cause injury by fumes in or to the territory of another, the properties or persons therein, when the case is of serious consequence and the injury is established by clear and convincing evidence'. Principle 21 of the Stockholm Declaration on the Human Environment of 1972 stated that 'states have, in accordance with the Charter of the United Nations and the principles of international law, the sovereign right to exploit their own resources pursuant to their own environmental policies, and the responsibility to ensure that activities within their jurisdiction or control do not cause damage to the environment of other States or of areas beyond the limits of national jurisdiction'. This principle was reaffirmed with some modifications in 1992 in the Rio Declaration. This principle is recognized as part of customary international law and provides the 'legal basis for bringing claims under customary law asserting liability for international damage' (Sands, 1995, pp. 190–94). The problem is understanding what constitutes environmental harm and how does one define causality? This is specially complicated in the case of climate change where cause and effect are difficult to determine.

Another principle is the polluter pays principle which calls on the costs of reducing pollution and dealing with the impacts of pollution on the polluter. This was adopted in 1972 in the OECD and has since been adopted in a number of forums and is now also a part of European Union law. This principle too has been included in the Rio Declaration of 1992.

The third relevant principle is that action should be taken on the basis of 'common but differentiated responsibilities' and respective capabilities in dealing with the climate change problem (see Article 3 of the FCCC and Principle 7 of the Rio Declaration). This principle states that countries should take action in accordance with their contribution to the problem and their ability to actually deal with the problem.

Taken together these three principles of international law imply that the polluters should take action to reduce the damage caused to other in accordance to the level of pollution and the ability to take action. From this perspective, all polluting countries and in particular the developed countries have to take serious measures to reduce their emissions and thereby reduce the potential harm to other countries. This is a position held by many environmental lawyers, countries like Brazil (Brazilian Proposal, 1997) and some NGOs like the Centre for Science and Environment in New Delhi.

2.6 Synthesis

The climate change problem is per se uncertain. However, by identifying the contours of the discussion one can identify reasons for action or inaction, and reasons for allocating responsibilities. The problem may be defined in terms of protecting vulnerable systems, and since the risk to these is very high the uncertainty level is reduced and the justification for action is high. The problem may be defined in terms of cause and effect, in which case all polluters need to take action. The problem may be defined as one of equity and justice, and here the uncertainty about responsibility is low but the uncertainty about the course of action and the level of compensation is high. The problem may be defined as one of efficiency. Although the uncertainty about the impacts of climate change is high, the uncertainty about the marginal costs of greenhouse gas emissions is smaller. All polluters should reduce their emissions, and the carbon tax should be the same for everyone. The problem may be defined as a precautionary approach. Then, there is little reason for concern in the near future, but this is rising rapidly as time progresses. The problem may be defined in terms of a legal approach and then the rich, greenhouse gas-emitting countries are expected to take far-reaching measures to reduce their own emissions and to help the poorer countries adopt different technologies and adapt to the problem of climate change.

Although there appears to be reason enough to take substantive action, and although there is no shortage of substantive ideas, technology and inspiration in the OECD countries, nevertheless countries are reluctant to take action.

Perhaps this is because of the implications of greenhouse gas emission reduction policy. Schelling (1997) argues that the problem with the climate change issue is that it tends to revolve around the sharing of a trillion-dollar emission business and that the world is unlikely to sit down and rationally divide such high value rights in perpetuity. That is why Cooper (1998) argues that the commitments under the Kyoto Protocol will be impossible to implement because they raise the problem of international allocation of emission rights, and that the focus should instead be on policies and measures.

Table 2.1 Perspectives on climate change

Definition	Parameters	Uncertainty of climate change	Justification for action	By whom
Unique, valuable ecosystems and human habitats	Ecological and national security approach	Low	High	Those polluting
Distribution, justice	Human rights approach	High	High	Those polluting
Aggregate impacts	Efficiency approach	Medium	Low	All
Large-scale discontinuities	Precautionary approach	High but decreasing	Low but increasing	All
Cause and effect; Responsibility approach	Polluter Pays Principle; State responsibility	Low (responsibility) or high (compensation)	High, because you do not know the risks for your country	Those polluting, in accordance with pollution levels and ability to pay

3   OTHER REASONS TO REDUCE EMISSIONS

In addition to the above differences of perception regarding the seriousness of the climate change problem, there is the issue of additional benefits that can be achieved through climate policy. These can be summed up as follows:

- improving resource use through industrial transformation, reducing inefficiencies in the production, distribution and consumption systems, and reducing resource dependency and solid, liquid and gaseous wastes; and
- making effective linkages to other environmental, political and social problems and thereby securing double dividends.

3.1   Development and Industrial Transformation

The problem of climate change is potentially caused by emissions from the energy-generation and energy-consumption sectors. Oil is the main source of energy, followed by gas and coal. Somewhere in the second half of the 21st century we will run out of conventional oil, even if oil exploration and extraction technologies continue to improve at their current rates. We will run out of gas well before that (Nakicenovic and Swart, 2000). Well before the middle of the 22nd century there will be other reasons to reduce dependency on imports. Most of the oil imports come from politically and economically unstable areas such as the Middle East and former Soviet Union. For economic reasons, peace and security in the exporting countries becomes of prime concern for the importing countries as was evident from the Gulf War of the 1990s. It is further argued that in a business-as-usual scenario, as oil resources dwindle the price will increase and poorer countries will be unable to afford oil imports, perhaps leading to international political tension and enhanced security problems (Jung and Loske, 2000). So the energy sector needs a drastic overhaul. Because the energy sector is large, with huge capital investments, slow depreciation and long lead times for new technologies, the transformation of the energy sector should start soon. In fact, it has started already with the investments of major energy companies in alternative energy sources.

Energy policy is presently a far more dominant political issue than climate policy. Not only do developed countries spend large sums of money on energy policy, including the development of technologies for energy saving and renewables, but the developing countries also prioritize energy programmes. Energy issues are not necessarily less sensitive, but at least the issues are seen as critical domestic priorities that need to be addressed. A recent project investigated the feasibility of modernizing the electricity sector in China and India (Gupta et al., 2001). The project concluded, *inter alia*, that both countries are in the process of modernizing their electricity producing and consuming sectors. The driving force is not climate change, but liberalization and the need to compete with foreign producers and products in the domestic and international market; and local air pollution. This has led to a number of policies and laws in both countries since 1990. The research concludes, *inter alia*, that in the period 1990–2020, although electricity production increases by a factor of five in both countries, related greenhouse gas emissions increase only by a factor of four and three respectively in China and India.

However, the problem is that energy policy is intricately linked with national security, economic and technological policies. Countries may seek to use locally-available energy sources in preference to imported sources even if such domestic energy sources and technologies are more polluting, because they are cheaper; and because it reduces the reliance of the national economy on imports. Energy policy calls for a combination of measures on the supply and demand side. On the supply side, the short-term focus tends to be on fuel switching, moving from coal-based power plants and other coal use to oil, gas, nuclear, large hydroelectricity plants and other renewables such as solar, small hydro, biomass and wind. In the long run, there are essentially four alternatives to – finite! – conventional oil and gas:

* coal
* unconventional oil and gas (from tar sands and clathrates)
* nuclear power
* renewables (biomass, solar, small hydro, wind)

To operate at the current scale of oil and gas, each alternative requires a large research and development programme, large investments in new production and consumption technologies, investment in processes to transport and, if necessary, import the fuels, changes in the transmission and distribution network, large changes in market structures, large financial investments and potential obstruction from vested interests and large geopolitical changes. All alternatives are either proven or fast becoming proven technologies, and are able to meet energy demands (Nakicenovic and Swart, 2000). In that sense, the four alternatives are equivalent. The results, of course, are not.

The implications for climate change are quite different. Standard business-as-usual scenarios, such as the IPCC's IS92a, assume that coal is the alternative of choice (Nakicenovic and Swart, 2000). An energy system based on coal would imply large carbon-dioxide emissions, and either high emissions of other pollutants or large environmental protection costs. A switch from conventional to unconventional oil and gas would imply small carbon-dioxide emissions, and as yet unknown environmental consequences.

Nuclear and renewables produce very low emissions of greenhouse gases during the electricity-producing phase. However, both downstream and upstream of the electricity production phase, nuclear power produces a range of serious pollutants, including carbon dioxide. Nuclear power is currently discredited and is being phased out in several European countries. The International Atomic Energy Agency and the governments of several developed and developing countries (as opposed to the public) would not be averse to the development of nuclear power in developing countries. The problem of climate change was seen as an opportunity for a comeback by nuclear power. However, a recent political decision of the sixth Conference of the Parties to the climate convention concluded that countries should refrain from acquiring emission reduction credits from projects that invest in nuclear power.

At this moment, renewables appear to have the support of environmental

NGOs, and several large oil companies are also investing in renewables. However, the main disadvantage of renewables is that their application on a large scale is untested, and may hide nasty surprises.

The real relevance for climate policy is as follows. Even without the problem of climate change, our energy systems have to be overhauled in order to cope with the economic and political consequences of the shrinking resource base and because of the local environmental impacts of the fossil-fuel extraction and use sectors. All drastic changes are painful and expensive, so it would make more sense to try and incrementally reshape the system with strategic incentives. From such a perspective, climate change would be seen more as an energy-resource problem than a global environmental problem. Energy policy is routine business for most governments, while climate policy is still struggling with its role and mandate.

While there is considerable emphasis on the supply side, energy saving also offers great opportunities for emission reduction in both developed and developing countries. IPCC (1996, p. 268; cf. Weiszäcker et al., 1997; Sachs et al., 1998) reports that there are huge cost-effective energy conservation and energy efficiency options available which offer economic and environmental benefits in addition to reducing emissions. The existence of the no-regrets options depends on the degree of market failure (lack of investment, distorted investment, imperfect information) in a country and the ability to develop policies to deal with these issues. The respective sizes of these technical inefficiencies are a matter of great dispute (Hourcade et al., 2000), but it seems that there is at least some no-regret emission reduction potential in the developed countries, in other words, emission abatement that is worth doing anyway. Good housekeeping measures can, in themselves, reduce energy consumption considerably. The Asian Dilemma project shows that the maximum potential energy saving in relation to the business-as-usual scenario for China and India in the period 1990–2020 is 30 per cent which amounts to a reduction of about 45 per cent of greenhouse gases (Gupta et al., 2001).

The end-use options call for a change of attitude in a large number of sectors, but in the long term are financially quite viable especially for the industries and companies that would like to survive the tough international competition. Investments in energy-efficient equipment and policies save on energy use, and hence resource use, and lead to savings on the part of individual consumers and of the generators of electricity. The small-scale sector will, however, face many difficulties. Such investments also reduce the need to import energy from other countries, which costs money, usually has security implications and, for developing countries, foreign exchange implications as well. From this perspective, there are sufficient economic and security reasons to explore alternative forms of energy and thereby reduce dependence on imports from or via unstable or unfriendly countries. A shift away from fossil fuel would change the face of the geopolitical situation. Obviously, some would benefit from this while others would lose. However, countries currently heavily dependent on imported energy are likely to be amongst the winners (Jung and Loske, 2000).

The need to use resources sparingly and efficiently and the process of liberalization in the developing countries therefore provides the motivation for investing in energy-efficient technologies in the supply and end-use sectors; and these investments will benefit the local producers and consumers. At the same time, this will reduce the pressure to quickly generate large amounts of new energy sources, giving time for countries to invest in those renewables in which they are rich and thereby address the problems of energy security and foreign accounts. These should be powerful incentives for change in the future. Careful choices in environmental policy could lead to substantial greenhouse gas emission reduction without an explicit climate policy. The same holds for transport problems. Greater emphasis on public transport and counteracting urban sprawl would reduce congestion and greenhouse gas emissions and improve urban air quality while saving on petrol imports and foreign exchange.

3.2 Other Issue Linkages

Apart from the benefits of industrial transformation of the energy sector, there are a number of reasons to take action in the area of greenhouse gases. Issue linkages can be used to make interests of diverging countries converge. Developed countries face the problem of stranded resources if they shift existing electricity-generation processes to renewable sources. Developing countries, by definition, need more generation facilities. However, at this moment renewable energy technologies are much more expensive and relatively less affordable for the developing countries. Co-operative approaches could be developed by groups of countries to match these two problems and help provide renewable technologies to developing countries at affordable prices. This could be done through, for example, the Clean Development Mechanism. Timber exporters and importers could try and make agreements to develop common eco-label schemes and ensure that the price of the timber is adequate for sustainable forestry to reduce the incentive for deforestation.

Issue linkages can be used to link domestic with global priorities. The developed countries need markets for their technologies and products. The developing countries would like to deal with local air and water pollution. This can in theory lead to fruitful cooperation.

Issue linkages can also be used domestically to prioritize policies that also support climate change. Biodiversity can best be preserved by maintaining and nourishing existing biodiversity reserves, and this can provide a powerful incentive for fighting deforestation, especially since biodiversity is seen as a key resource of the South. Desertification can, in theory, be countered by growing appropriate vegetation that can lead to the development of sinks, reduce soil erosion and may lead to other environmental, social and economic benefits. Methane from landfills can be collected and used as a source of energy.

Another example is the carbon tax. Carbon taxes would generate welcome revenue for governments. In fact, carbon taxes would fit in with the current tendency of OECD governments to replace income-based taxes with consumption- and pollution-based taxes. This may have undesirable effects on income distribution, since income taxes are progressive while consumption taxes are regressive. However, tax reform may well have positive effects on productivity and unemployment, particularly in the high-labour-cost countries of Continental Europe.

4   DILEMMAS FACING COUNTRIES

There are therefore enough reasons for countries to take action. So why is there so much hesitation? One could identify the reasons in terms of dilemmas facing the developed and the developing countries.

4.1   Developed Countries

The developed countries face a number of social, political and economic dilemmas in choosing between alternatives. In the climate change issue, a key choice faced by these countries is whether they should reduce emissions now or whether it would be better to focus on technological progress and develop the technologies for reducing emissions in the future. Another key dilemma for the developed countries is how to adopt industrial transformation without sacrificing economic growth. While industrial transformation in the long term is likely to benefit individual countries, in the short term the country that takes domestic measures and the company that adopts new policies may find that their products are no longer able to compete in the international market. This leads to the development of special policies to exempt the energy-intensive sectors from carbon taxes, or to the development of policies that until other countries adopt similar measures, individual countries and companies will not take far-reaching steps. A consequence of adopting alternative technologies, production processes and/or electricity-generation processes is that existing technologies and processes become obsolete and worthless even though they are not at the end of their productive cycle. This leads to the problem of squandering resources by stranding them. Is this the most efficient use of the resources? Should such technologies and processes then be made available to developing countries? This is another dilemma faced by developed countries (cf. Maurer, 2000; Metz et al., 2001). A further issue is that while it is evident that the developed countries need to make some resources available to the developing countries to help them cope with the problem of climate change, most countries (developed or developing) would like to shy away from taking responsibility under the 'polluter pays' principle and prefer not to talk in terms of liability and compensation. This leads to the development of aid and trade mechanisms, which brings us to the next dilemma: whether the private sector, with its focus on profits, is interested in addressing global environmental problems or in short-term solutions that appear to be addressing global environmental problems (CEO, 2000). Another interesting dilemma is that, although human rights issues are key concerns for the developed countries, in the area of environmental negotiations human rights appear to be taking a back seat to property rights claims in relation to the atmosphere. This dilemma is seen in the discussion regarding how property rights to the atmosphere are to be shared between countries (cf. Schelling, 1997; Meyer, 2000; Brazilian Proposal, 1997). Finally, while developed countries would like to engage in productive negotiations that address the problem that they have highlighted, they also want to ensure that the solution costs their individual country the least and does not lead to over-commitment. The balance between showing leadership and protecting domestic interests is an interesting struggle to watch (cf. Ott, 2001; Grubb et al., 1999; Gupta and Grubb, 2000), especially in light of President George W. Bush's recent statement withdrawing support for the Kyoto Protocol. These dilemmas are summarized in Table 2.2.

Table 2.2   Dilemmas for the developed countries

Dilemma	Description	Relation to climate change
Timing	Taking action now or later	Is early emission reduction required or should current action focus on research and development, and regime building?
Industrial transformation	Transformation without sacrificing growth	How to adopt new technologies without running the risk of losing industrial competitiveness.
Wealth – 1	Spending without squandering	How to deal with 'stranded' resources; can one phase out power stations that still have generation potential without losing capital.
Wealth – 2	Assisting without compensating	Clearly there is a responsibility for those who have emitted the most and those who have the greatest ability to take action. The issue then is how to assist other countries on the basis of this concept, without being bogged down in discussions about compensation.
Privatization	Empowering private sector to solve public problems	The governments fear that the international problem cannot be addressed by governments alone, and so they are transferring responsibility to those very industries that caused the problem in the first place.
Ecospace	Human rights or property rights	The governments in developed countries that have fought hard to promote the concepts of human rights are now, instead, using the concept of property rights to share the available environmental space.
Negotiation	Negotiate pragmatically without committing too much	Developed countries have invested huge amounts of financial and human resources to the international negotiation process, but they are afraid to commit too much.

4.2   Developing Countries

Many developing countries claim that they have the right to development. (This claim is not unique to them since many developed countries also claim this right, although not quite in those terms.) This right is used to justify the continued use of polluting industries and equipment. Some argue that, being poorer, they have no choice but to use out-of-date equipment, and locally-available resources. Taking loans in the international market to finance modern technologies may only exacerbate the problem of international debt. The goods and services generated may therefore not be easily accessible and affordable to the people in the country and thus may need to be subsidized. While there is nothing intrinsically wrong with this argument, it is often used without being evaluated in the context of domestic options in different sectors. The lack of creative thinking in developing countries about development without pollution is a problem that developing countries may drown in. As an articulate interviewee once said:

> The government of India does not know the interests of the people of India. They represent elite interests of a hierarchical society. They represent interests that are similar to those of the North. If they were interested they could focus on the interests of the poor. They want to argue that you guys are causing so much pollution so we should also have the right to do so; it is a silly argument although understandable. It is based on the comparative perspective and based on a competitive framework of what is right and fair. If you take a more balanced and holistic view, you realise that the argument goes nowhere. It's just not feasible. You cannot continuously increase consumption and there are distinctions between different forms of consumption, between necessities, reckless obsolescence and privatisation of activities that should be in a public sphere. We need an alternative model of development and alternative approaches to climate change. We need to stop this emulative madness. India could make certain commitments on GHG reductions – it should push for considerable reductions through producer liability for emissions and waste disposal; and take a proactive approach and form strong networks with the other DCs.

However, as many other interviewees have pointed out, some developing countries aspire to be like the West and are not engaged seriously in discussions of sustainability and security. The dilemma can be summed up as follows: how does one define 'modernization without westernization'? This dilemma is scarcely articulated in the South Commission Report (1990) and there were many critics of the 'copycat' inclinations in the report. 'On the contrary, it would seem as though the development of the South implied its integration – almost copycat integration – into the North-dominated system' (Comeliau, 1993, p. 68; Kothari, 1993, p. 85; Pisani, 1993). Its focus was too macro-economic, state-oriented, catching-up, and it called for a new international economic order that is similar to the old order except in the sharing of power (Comeliau, 1993, pp. 65–71). The search for a way to approach the North–South dialogue outside the boundaries of existing political discourse is thus evidenced by the literature on North–South issues.

One can identify six dilemma's faced by developing countries. These are shown in Table 2.3.

Table 2.3   Dilemmas for the developing countries

The dilemmas	Description	Issues
The development dilemma	How to modernize without westernizing?	Climate change and Agenda 21.
The poverty dilemma – I	How to survive without squandering one's resources?	CITES, low-hanging fruit in the Clean Development Mechanism; the Basel Convention; the Biodiversity issue.
The poverty dilemma – II	How to beg without mortgaging one's future?	Agenda 21; Desertification; GEF.
The privatization dilemma – I	How to empower the private sector to solve public problems?	Climate change; Biodiversity.
The environmental space dilemma	How to achieve equity in one regime without being held responsible in another area/level?	Climate change; biodiversity; domestic policy.
The economic dilemma	How to serve short-term business interests without affecting long-term economic interests?	Trade; debt; eco-labelling.

Source: Gupta (2002).

Clearly some developing countries face these dilemmas more than others. All developing countries with extreme divergence in income and lifestyles face these dilemmas. A handful of countries with very small populations and homogeneous structures that are at risk from climate change may not face these dilemmas.

5   CONCLUSIONS

With the publication of the first IPCC reports, countries adopted a positive and determined approach to dealing with the climate change issue. As the regime is developing over time, there is increasing reluctance on the part of countries to take on serious obligations. This is very much in line with the predictions Downs (1972) made in his theory of the issue–attention cycle. Downs argues that in the initial stages of a problem, with the growth of public support, policymakers are enthusiastic and believe that if they put their mind to it the problem can be solved. If policies are adopted without hesitation at this stage, the problem may be controlled. However, as time passes, if it appears that the costs of dealing with the problem may be quite high, the problem gradually fades out of the public limelight and securing the political commitment to take action becomes difficult.

The climate change problem is not likely to disappear from the agenda. There will be increasing evidence of the problem over the coming years and the need to take action. However, given the inherent uncertainties in the problem, and the enormous stakes involved, the decision to support action will depend on the perspective of the stakeholders concerned. For the majority of natural scientists, ecologists, economists and other researchers associated with the climate change process, there is already enough evidence of the need to take action. For the potential victims of climate change the evidence can only become more convincing at a serious cost to them. However, it is uncertain whether these groups of actors are powerful enough to force change. There are many sceptics, as well as people who would lose from emission abatement and have a larger financial stake.

At the same time, there are plenty of other reasons to justify action. Policies that focus on energy efficiency lead to a reduction of demand for energy generation, save natural resources, reduce pollution upstream and downstream of the energy-generation process, reduce the energy bill for individual consumers and producers, reduce the costs of fossil-fuel imports and thereby reduce foreign-exchange depletion and increase the energy security of the nation. Such policies may also lead to ancillary benefits and reduce deforestation, land degradation and desertification, local air pollution and related health issues, and so on. This is not just a powerful argument for the developed countries, but is becoming an increasingly powerful argument in the developing countries as well.

At the same time, making trade-offs between different options is not easy. The dilemmas in adopting policies to reduce emissions of greenhouse gases at both developed and developing country level and the uncertainty of the science thus become major hurdles for policy development and implementation.

NOTES

1. Richard Tol acknowledges financial support by the US National Science Foundation through the Center for Integrated Study of the Human Dimensions of Global Change (SBR-9521914), and the Michael Otto Foundation. Joyeeta Gupta acknowledges the financial support from the project: Climate Change and the Law of Sustainable Development, a USF project of the Vrije Universiteit Amsterdam.
2. The negotiations include 184 countries and there are another 10 countries that are observers or not participating.
3. There are annual negotiations of two weeks and preparatory negotiations of between six and eight weeks.

REFERENCES

Böttcher, F. and J. Emsley (eds) (1996), *Global Warming: A Report from the European Science Foundation*, London: European Science Foundation.

Brazilian Proposal (1997), *Proposed Elements of a Protocol*, FCCC/AGBM/1997/Misc.1/Add.3.

CEO (2000), *Greenhouse Market Mania: UN Climate Talks Corrupted by Corporate Pseudo Solutions*, Amsterdam: Corporate Europe Observatory.

Comeliau, C. (1993), 'The South: Global Challenges', in South Centre (ed.), *Facing the Challenge: Responses to the Report of the South Commission*, UK: Zed Books, pp. 67–75.

Cooper, R.N. (1998), 'Toward a Real Global Warming Treaty', *Foreign Affairs*, **77**(2), 66–79.

Downs, A. (1972), 'Up And Down With Ecology – The Issue–Attention Cycle', *The Public Interest*, **28**, 38–50.

Grubb, M. and F. Yamin (2001), 'Climatic Collapse at The Hague: What Happened, Why and Where Do We Go From Here?', *International Affairs*, **77**(2), 261–76.

Grubb, M., C. Vrolijk and D. Brack (1999), *The Kyoto Protocol*, London, UK: Earthscan/RIIA.

Gupta, J. and M. Grubb (eds) (2000), *Climate Change and European Leadership: A Sustainable Role for Europe*, Environment and Policy

Series, Dordrecht, The Netherlands: Kluwer Academic Publishers.

Gupta, J. (2002), 'Environment and Development: Towards a Fair Distribution of Burdens and Benefits', in J.J.F. Heins and G.D. Thijs (eds), *Ontwikkelingsproblematiek: The Winner Takes It All*, Themabundel Ontwikkelingsproblematiek Nr. 12, p.35-50, Netherlands: Vrije Universiteit Amsterdam Press.

Gupta, J., J. Vlasblom, C. Kroeze, with Kornelis Blok, Jan-Willem Bode, Christiaan Boudri, Kees Dorland and Matthijs Hisschemoller (2001), *An Asian Dilemma: Modernising the Electricity Sector in China and India in the Context of Rapid Economic Growth and the Concern for Climate Change, Institute for Environmental Studies*, Report of the Netherlands National Programme on Global Air Pollution and Climate Change, Bilthoven, The Netherlands: NOP.

Houghton, J.T., L.G. Meira Filho, B.A. Callander, N. Harris, A. Kattenberg and K. Maskell (eds) (1996), *Climate Change 1995: The Science of Climate Change*, Cambridge, UK: Cambridge University Press.

Hourcade, J.C., P. Courtois and T. Lepesant (2000), 'Socio-Economics of Policy Formation and Choices', in J. Gupta and M. Grubb (eds), *Climate Change and European Leadership: A Sustainable Role for Europe*, Netherlands: Kluwer Academic Publishers, pp. 109–34.

IPCC (1996), *Climate Change 1995: Economic and Social Dimensions of Climate Change*, Cambridge, UK: Cambridge University Press.

IPCC (2001), *Climate Change 2001 – Mitigation*, Cambridge, UK: Cambridge University Press.

IPCC-I (2001), *Climate Change 2001: The Scientific Basis*, Cambridge, UK: Cambridge University Press.

Jung, W. and R. Loske (2000), 'Issue Linkages to the Sustainability Agenda', in J. Gupta and M. Grubb (eds), *Climate Change and European Leadership: A Sustainable Role for Europe*, Environment and Policy Series, Netherlands: Kluwer Academic Publishers, pp. 157–72.

Kothari, R. (1993), 'Towards a Politics of the South', in South Centre (ed.). *Facing the Challenge; Responses to the Report of the South Commission*, UK: Zed Books, pp. 84–91.

Maurer, C. (2000), 'Rich Nations' Investments Heighten Climate Risk', *International Herald Tribune*, 18–19 November, 2000, p. 14.

Metz, B., O.R. Davidson, J.-W. Martens, S.N.M. van Rooijen and L.V.W. McGrogy (eds) (2001), *Methodological and Technological Issues in Technology Transfer*, Inter Governmental Panel on Climate Change, Cambridge, UK: Cambridge University Press.

Meyer, A. (2000), *Contraction and Convergence: The Global Solution to Climate Change*, Schumacher Briefings, No. 5, Foxhole, Darlington,

Totnes, Devon, UK: Green Books for the Schumacher Society.

Nakicenovic, N. and R. Swart (eds) (2000), *Emissions Scenarios 2000*, Special Report of the Intergovernmental Panel on Climate Change, Cambridge, UK: Cambridge University Press.

Oberthür, S. and H.E. Ott (1999), *The Kyoto Protocol. International Climate Policy for the 21st Century*, Berlin, Heidelberg et al., Germany: Springer Verlag.

Ott, H.E. (2001), 'Climate Change: An Important Foreign Policy Issue', *International Affairs*, **77**(2), 277–96.

Pisani, E. (1993), 'Inventing the Future', in South Centre (ed.), *Facing the Challenge; Responses to the Report of the South Commission*, UK: Zed Books, pp. 97–9.

Sands, P. (1995), *Principles of International Environmental Law, Vol. 1. Frameworks, Standards and Implementation*, Manchester, UK and New York, US: Manchester University Press.

Sachs, W., R. Loske and M. Linz et al. (1998), *Greening the North: A Post-Industrial Blue Print for Ecology and Equity*, London, UK: Zed Books.

Schelling, Thomas C. (1997), 'The Cost of Combating Global Warming: Facing the Trade-Offs', *Foreign Affairs*, **76**(6), 8–14.

Sluijs, Jeroen P. Van der (1997), *Anchoring Amid Uncertainty; On the Management of Uncertainties in Risk Assessment of Anthropogenic Climate Change*, University of Utrecht, Utrecht, The Netherlands (ISBN 90-393-1329-6).

Smith, J.B., H.-J. Schellnhuber, M.M.Q. Mirza, S. Fankhauser, R. Leemans, E.Lin, L. Ogallo, B. Pittock, R.G. Richels, C. Rosenzweig, R.S.J. Tol, J.P. Weyant and G.W. Yohe (2001), 'Vulnerability to Climate Change and Reasons for Concern: A Synthesis', in J.J. McCarthy, O.F. Canziani, N.A. Leary, D.J. Dokken and K.S. White (eds), *Climate Change 2001: Impacts, Adaptation, and Vulnerability*, Cambridge, UK: Cambridge University Press, pp. 913–67.

South Centre (1993), 'An Overview and Summary of the Report of the South Commission', in South Centre (ed.), *Facing the Challenge: Responses to the Report of the South Commission*, UK: Zed Books, pp. 3–52.

Weiszäcker, E. Von-, A. Lovins and H. Lovins (1997), *Factor Four, Doubling Wealth and Halving Resource Use*, London, UK: Earthscan.

3. Challenges of future climate policy: what can be expected?

Ton Manders and Richard S.J. Tol

1 INTRODUCTION

This chapter sketches some of the challenges for climate policy that lay ahead. Options for mitigation of climate change depend on our visions of the future. Hence, dealing with a long-term issues like climate change calls for insight into that future. It is an understatement to say that our future is uncertain. Scenario building is a way of assessing this uncertainty, but there is no single unique way to proceed. Scenarios appear in many guises, ranging from pseudo-exact forecasts to compelling stories (whether fairy tales or nightmares).

Emissions of greenhouse gases are the ultimate driving force behind climate change. In this chapter, we explore some of the more recent emissions scenarios, addressing both the methodology and the results. In light of these scenarios and the associated impacts of climate change, we discuss whether action should be taken and what the implications of a cost-effective mitigation strategy are. Given the wish to stabilize the concentration of greenhouse gases in the atmosphere, we discuss and illustrate some of the issues that arise. This chapter relies on recent work initiated by the Intergovernmental Panel on Climate Change (IPCC) (IPCC, 2000, and IPCC, 2001a, b, c).

This chapter starts with some preliminary remarks on scenario building (Section 2). We then focus on the new emissions scenarios relevant for the climate change debate (Section 3). On the basis of these projections and their possible implications for the climate, we discuss five reasons to abate emissions (Section 4). We then consider the limitations and consequences of a cost–benefit analysis, a commonly used method to investigate abatement strategies (Section 5). Given the desire for abatement, we explore a number of stabilization scenarios and discuss some of the issues that they raise (Section 6). Finally, we present some concluding remarks (Section 7).

2   MANY APPROACHES TO THE FUTURE

Scenarios are a useful and fairly common method used by policymakers to think about an uncertain future and to review policy actions that have long-term effects or that may gain in importance over time. Scenarios are particularly useful when uncertainty surrounds not only the realization of stochastic variables like economic growth or investment, but also current and future characteristics of entire systems, like the economy or the environment.

Perhaps it is useful to make a distinction between two types of scenarios. The first is a scenario that typically starts from 'what if' and goes on to describe extreme, even somewhat implausible, situations in the future. This type may have the advantage that it forces readers to consider a world that is not a mere reproduction and extrapolation of the current situation and thus forces them to 'think the unthinkable'. The second type is a scenario that provides a coherent background to assess various policy actions and combines several trends. However, to try every combination of trends is neither feasible nor optimal. There are simply too many combinations. The art to preparing a scenario, therefore, is to make a few intelligent combinations of trends.

One should differentiate clearly between policy scenarios and no-policy scenarios. Scenarios are often used to sketch the policy challenges that will emerge sooner or later. It seems logical to exclude from the scenario those actions that are a possible response to these challenges. In this way, the consequences of various policy actions for these problems can be determined and assessed. One disadvantage is that the scenario may become somewhat implausible. Would governments really not respond if they see problems emerging? Besides, policy actions may give rise to new problems and thus change the challenges that current and future policymakers may have to deal with. Therefore, an approach in which a scenario includes policy reactions is sometimes also useful. Both approaches – to keep policy actions exogenous or to endogenize policy reactions – have their merits. However, most scenarios are intended for discussions about future policy, and the approach of exogenous policy actions in at least that policy domain is more adequate.

In general, there are two stereotypical approaches to creating scenarios: the narrative approach and the modelling approach. In the narrative approach, a scenario is mainly a story. This approach has the advantage of providing a high degree of freedom to construct a scenario. Such a scenario can be a source of inspiration. But the freedom to imagine is also a disadvantage. There are several dangers in a story: it may overemphasize a mechanism, it may inadvertently change causality and it may overlook relations and effects. This is likely to undermine the credibility of the story. The modelling approach does not normally suffer from these disadvantages, and has a number of strengths. It draws upon current insights into (economic) theory and empiricism. This validates the outcome of the model. A scenario must also take into account straightforward but essential constraints, whether physical, economic or statistical. A model explicitly defines constraints and cannot 'forget' them. This guarantees internal consistency and gives the outcome credibility. Furthermore, this approach allows one to describe not only the situation at the end of the scenario period but also the entire path that leads to the end situation. This makes future situations consistent with the current situation and gives more plausibility to the scenario. In actual practice, scenario builders merge both narrative and quantitative elements (see CPB, 1992 and IPCC, 2000).

3 EMISSIONS SCENARIOS

We now turn to scenarios relevant to climate change policy. Policies can only be discussed meaningfully if we first answer the question: what might happen if we do nothing? Projections of the impact of climate change on societies depend to a large extent on projections of the temperature change in the future. Temperature change is closely related to the concentration of greenhouse gases in the atmosphere, which in turn depends on earlier emissions. So ultimately, emissions drive the climate effects. Emissions depend on population growth, economic development and technological progress.

The IPCC has developed a number of new emissions scenarios in the Special Report on Emissions Scenarios (IPCC, 2000). These so-called SRES scenarios are designed to be non-mitigation or reference scenarios, in other words, scenarios in which additional policy initiatives aimed specifically at reducing greenhouse gas emissions are absent. This section describes the SRES scenarios.

3.1 Methodology of SRES

In the SRES process, a qualitative, narrative approach was fused with a more formal approach using different models to guarantee structural variance and methodological diversity in the scenarios. Consequently, the SRES scenarios combine elements of both the more story-like scenarios and the more model-based scenarios. The SRES process also sought to improve scenario development by:

- extensively documenting the inputs and assumptions of the SRES scenarios;
- formulating storylines;
- encouraging a diversity of approaches and methods for deriving scenarios;
- making the scenarios from different groups more comparable;
- assessing the differences and similarities between those scenarios;
- considering a wider range of economic development pathways, including a narrowing of the income gap between developing and industrially-developed countries;
- incorporating the latest information on economic restructuring throughout the world;
- examining different trends in and current rates of technological change.

As a result, SRES made a big step forward in the methodology of scenario building. Nevertheless, some problems remain, particularly with regard to consistency with observations. For instance, in some scenarios China is projected to continue to grow at breakneck speed for another 30 years. In most scenarios, worldwide income disparities are assumed to decline even though they have been widening for the last 50 years.

The basic approach involved the formulation of the qualitative characteristics of a scenario in the form of storylines, and then quantification using different modelling approaches. The qualitative description gives background information about the global setting of the scenarios, which can be used to assess society's capacity to adapt to and mitigate climate change and to link the emissions scenarios with issues of development, equity and sustainability. The quantitative description of emissions scenarios can be used as input for models to compute the future extent of climate change and to assess strategies to reduce emissions.

Four scenario 'families' were developed. An even number helps to avoid the impression that there is a 'central' or 'most likely' case. The scenarios cover a wide range of possible futures, but not all. In particular, there are no global or even regional disaster scenarios. None of the scenarios include explicit new climate policies. Each family has a unifying theme in the form of a 'storyline', or narrative, that describes future demographic, social, economic, technological and policy trends. The quantification involved first translating the storylines into a set of quantitative assumptions about the driving forces of emissions (for example rates of change of population and size of the economy and rates of technological change). These assumptions were then used as input for models that computed the emissions of greenhouse gases. The result was a total of 40 scenarios for the four storylines. The large number of alternative scenarios showed that a single storyline could lead to a large number of feasible emission pathways.

Since by definition there is no agreement on how the future will unfold, the SRES tried to bring the alternatives into sharper focus by assuming that scenarios have diverging tendencies. A scenario either emphasizes stronger economic values or stronger environmental values; a scenario either assumes increasing globalization or increasing regionalization. Combining these tendencies yielded the four different scenario families (see Figure 3.1). The two-dimensional representation of the main SRES scenario characteristics in Figure 3.1 is an over-simplification. It is merely an illustration. To be accurate, the space would in fact need to be multi-dimensional, listing other scenario developments in many different social, economic, technological, environmental and policy dimensions.

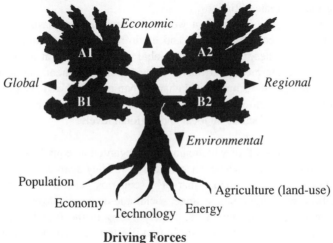

Driving Forces

Source: IPCC (2001c).

Figure 3.1   Illustration of SRES scenarios

The names of the scenario storylines and families, A1, A2, B1, and B2, are purposely unimaginative:

- The A1 storyline describes a future world of very rapid economic growth, population that peaks in mid-century and declines thereafter, and the rapid introduction of new and more efficient technologies. Major underlying themes are convergence among regions and increased cultural and social interaction, with a substantial reduction in regional differences in per capita income.

- The A2 storyline and scenario family describe a very heterogeneous world. The underlying theme is self-reliance and preservation of local identities. Fertility patterns across regions converge very slowly, which results in a steadily increasing world population. Economic development is primarily regionally oriented and per capita economic growth and technological change are more fragmented and slower than in other storylines.
- The B1 storyline describes a convergent world with the same world population as in the A1 storyline, but with rapid changes in economic structures toward a service and information economy, with reductions in material intensity, and the introduction of clean and resource-efficient technologies. The emphasis is on global solutions to economic, social, and environmental sustainability, including greater equity, but without additional climate initiatives.
- The B2 storyline describes a world in which the emphasis is on local solutions to economic, social, and environmental sustainability. It is a world with a steadily increasing global population but at a slower rate than in A2, intermediate levels of economic development, and less rapid and more diverse technological change than in the B1 and A1 storylines. While the scenario is also oriented toward environmental protection and social equity, it focuses on local and regional levels.

Six of the scenarios, all of which should be regarded as equally sound, were chosen to illustrate the whole set of scenarios. They span a wide range of uncertainty. Four so-called 'marker' scenarios encompass the four combinations of demographic change, social and economic development and broad technological developments, corresponding to the four families (A1, A2, B1, and B2). From the A1 family two illustrative scenarios were added (A1FI, A1T). They explicitly explore alternative energy-technology developments, holding the other driving forces constant: fossil-intensive (A1FI), non-fossil energy sources (A1T) or a balance across all sources (the 'marker' A1B). Rapid economic growth leads to high capital turnover rates, which means that minor early differences between scenarios can lead to a large divergence by 2100. The A1 family, which has the highest rates of technological change and economic development, was therefore selected to show this effect.

3.2 Basic findings of SRES

Figure 3.2 illustrates the range of global energy-related and industrial carbon-dioxide emissions for the six scenarios.

(1990 = 100)

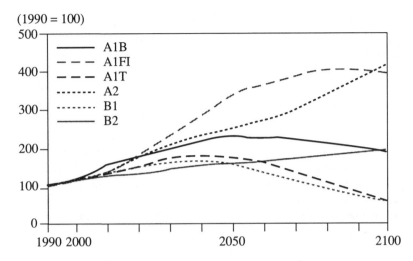

Figure 3.2   Global CO$_2$ emissions in six SRES scenarios from 1990 to 2100 (1990 = 1)

Figure 3.2 shows that the four 'marker' and two illustrative scenarios in themselves cover a large portion of possible distributions. This is one reason for the use of all six scenarios in future assessments. Together, they cover most of the uncertainty about future emissions.

The SRES scenarios lead to the following conclusions:

- Alternative combinations of driving-force variables can lead to similar levels of emissions. For instance, significant global changes could result from both a scenario of high population growth where per capita incomes rise only modestly and a scenario in which a rapid demographic transition (low population levels) coincides with high rates of income growth.
- Within the same scenario family future developments may diverge strongly (for example A1).
- There is no unique ordering of scenarios. Emission profiles from different scenarios may intersect. Rising trends may reverse in the second half of this century.
- There is simply no question of one, and only one, possible development path (as alluded to for instance in concepts such as the 'business-as-usual scenario'). There are numerous combinations of driving forces and numerical values that can be consistent with a particular scenario description.

All of the SRES marker scenarios lead to higher concentrations of greenhouse gases in the future. Only the 'low emission' scenario B1 leads to stabilization by the end of the century, although at a concentration level that is more than twice the pre-industrial level. Other scenarios result in even higher levels in 2100 and show concentrations continuing to rise in the 22nd century. It is becoming increasingly clear that a higher greenhouse gas concentration leads to a warmer and wetter world (IPCC, 2001a). Over the 20th century the temperature has risen by about 0.6°C and SRES scenarios project a further increase by 1.5 to 6°C by 2100. Global mean sea level is projected to rise by 0.10 to 0.88 metres between 1990 and 2100. Global warming is expected to bring more frequent and more intense summer heat-waves, warmer winters and an increased risk of winter floods. The effects will not be spread evenly over the earth (IPCC, 2001a, 2001b).

4   FIVE REASONS TO ABATE EMISSIONS

People have a variety of reasons to be worried about climate change and for wanting to reduce greenhouse gas emissions. Gupta and Tol (chapter 2, above) offer a classification of reasons for concern, largely based on the Third Assessment Report of IPCC WG2 out (IPCC, 2001b). This classification gives different reasons to abate emissions:

1. Unique and vulnerable systems
2. Equity and distribution
3. Efficiency and economic development
4. Large-scale disruption
5. Responsibility

These reasons represent different viewpoints. The alternatives are named after the aspect of the impact they emphasize. We briefly summarize the five reasons here.

Climate change will affect everyone and everything, but some people and places are more vulnerable than others. This includes ecosystems, species and people with limited possibilities to adapt, either because they are restricted geographically (islands, mountains) or because they are adapted to specific eco-climatic niches. Climate change may well cause substantial damage to the marginalized and the specialized. To some, the impact of climate change on 'unique and vulnerable systems' is itself sufficient reason to reduce greenhouse gas emissions. If this is indeed one's frame of mind, then climate change is very worrying, because such impacts have already been observed in relation to recent warming.

There is a strong negative correlation between greenhouse gas emissions and vulnerability to climate change. The rich countries emit most, while the poor countries will suffer most of the damage. Climate change may thus widen 'the gap between rich and poor'. To some, this is a reason for concern. Impact studies show that such people are right to be to be worried about climate change, although reducing greenhouse gas emissions is not necessarily the logical implication. Poverty alleviation may be a more (cost-) effective strategy, while emission abatement may have negative implications for the economic growth of poor countries (Tol and Dowlatabadi, forthcoming).

Some view climate change as an 'efficiency problem', and are principally concerned about its aggregate impact and its effect on economic development. This group should not be overly concerned because the exposure of the economy to weather and climate is small and shrinking and many negative effects are cancelled out by positive impacts (IPCC, 2001b). Developing countries are more vulnerable, but their economic weight is small. This is not to say that climate change should not be regarded as an efficiency problem. Indeed, studies suggest that the marginal cost of carbon-dioxide emissions (and, hence, the appropriate tax on emissions to restore this externality) is around \$20 per tonne of carbon (Tol, 1999c).

Yet others argue that the main reason to be concerned about climate change is the potential for 'large-scale disruptions', such as a break-up of the West-Antarctic Ice Sheet or a slowdown of the thermohaline circulation. The former would lead to a rise in the sea level by metres, while the latter could mean rapid cooling in Western Europe (and additional warming in Africa). Although possibly harmful, the probability of such events is currently unknown, while there has been scarcely any study of their potential impact. Citing the precautionary principle, some argue that this is a reason for emission reduction. It is also a reason for further research.

Finally, one could adopt 'a legal or moral perspective'. Since emissions cause harm to many, one should avoid such emissions or offer compensation (Gupta and Tol, Chapter 2, above). A strict interpretation of the 'polluter pays' principle would be very worrisome to large emitters, as climate change is expected to kill hundreds of thousands of people, force many more to migrate and yet more to change their way of life (Kemfert and Tol, 2001). However, international law is too weak to enforce this principle, while morality also tends to play a limited role in international politics (although less so in the accompanying rhetoric).

5 EFFICIENT SOLUTIONS

Some policy studies use cost–benefit analyses to investigate international greenhouse gas emission reduction strategies. However, one should take great care in interpreting the results. To make the analysis operational, a large number of simplifying assumptions has to be made. To evaluate the impact of greenhouse gas emission reduction and of climate change, the analyst is often forced to explore the limits of our knowledge. In a cost-benefit analysis, decisions with (implicit) ethical implications also have to be made about what counts and what does not, and about how things are compared and aggregated. Nonetheless, it is often helpful to consider (Pareto-) efficient policies and to ignore distributional considerations. The result can be used as a yardstick to gauge more realistic policy strategies.

Following pioneering work by Nordhaus (1992), the usual method of cost–benefit analysis is to use an extended growth model. Growth models are used to analyse the optimal savings rate. For climate change purposes, growth models are typically extended with greenhouse gas emissions, a simple model of atmospheric chemistry and climate and the impact of climate change on production, consumption, and welfare.

Extensive analyses of such models have led to the following conclusions (Nordhaus, 1994; Peck and Teisberg, 1994; Kolstad, 1998; Tol, 1999b). It is not in the best interest of mankind to stabilize atmospheric concentrations of greenhouse gases; on the contrary, welfare is maximized if concentrations are allowed to rise over time. The reasoning is as follows. If greenhouse gas emissions are reduced, climate change declines and occurs more slowly, so the impacts are smaller and the additional impact prevented by further emission reductions are smaller too. At the same time, the more emissions are reduced, the higher the costs of additional emission reduction. Thus, the costs of additional action rise, while its benefits fall. Reducing impacts to zero cannot be justified. So if climate change impacts depend largely on the rate of climate change (that is the enhanced greenhouse effect is a flow pollutant), there is no reason to stabilize atmospheric concentrations.

If, on the other hand, the impact depends largely on the level of climate change (that is the enhanced greenhouse effect is a stock pollutant), stable atmospheric concentrations would still imply sustained damage due to climate change. Emission reduction would, however, also continue to cost money as governments would need to keep emissions at a low level. It is impossible to say, without quantitative modelling, which of the two is more expensive. However, extensive model runs suggest that atmospheric stabilization, at any level, would cost more than it delivers. The reason is that running an economy with low greenhouse gas emissions is generally believed to be very expensive.

There are a few conditions under which this conclusion does not hold. First, there may be a cheap alternative to fossil-fuel combustion available on a sufficiently large scale, or perhaps an alternative whose real costs are hidden for a while (Peck and Teisberg, 1992). Second, the impact might suddenly become unbearably large if climate change were to pass a certain threshold (Manne and Richels, 1998). Impact research, so far, has failed to find such thresholds despite considerable efforts to do so (IPCC, 2001b). Third, it may be that the relative value of climate change impacts rises as fast as (or faster than) the discount rate (Hasselmann et al., 1997). Although theoretically possible, there is little or no empirical support for this claim (Kriström and Riera, 1996).

Not only is it hard to defend atmospheric stabilization from an efficiency viewpoint, it is also hard to defend the ambitious emission reduction targets (a few per cent per year) that OECD governments currently contemplate in international negotiations. If the costs of emission abatement are approximately what current studies find they would be, and if the costs of climate change are about the size that current studies suggest they would be, then optimal emission reduction is modest (less than 1 per cent per year).

This finding is reasonably robust to variations in parameters, including the discount rate, to alternative specification of impacts of climate change and costs of emission reduction, to uncertainties and to considerations of equity (Tol, 1999a, d, 2001). The finding is not robust to alternative decision criteria, such as the precautionary principle (Section 6, below). Of particular interest is the conclusion that, although policy recommendations for the long term may differ considerably between cases and studies, in the short term virtually all analyses point out that it would be advisable to start reducing greenhouse gas emissions at a slower pace than current policymakers deem necessary.

6　STABILIZATION SCENARIOS

To stop further global warming, concentrations of greenhouse gases should stabilize in the future. In most cases, this means determined action should be taken to curb emissions. What is needed to reach a meaningful stabilization target? That is the question we address in this section. The mitigation effort depends on the development of emissions in the reference scenario, the targets set and policies and measures taken to reach these targets. This section reflects on some of the post-SRES work (Morita, 2000). As a follow-up to the SRES process, a group of modellers quantified a number of stabilization scenarios using the IPCC's new SRES scenarios as a baseline. Some illustrations are taken from joint work by CPB and RIVM on stabilization, which was part of this post-SRES exercise (Bollen et al., 2000). When discussing stabilization scenarios, it should be noted that this is a partial analysis: how can we reach a given target, and what are the costs? This is not an integrated assessment where mitigation costs are balanced against the costs of climate change and where the optimal mitigation is determined. As Section 5 showed, an integrated assessment could lead to quite different conclusions.

Stabilization scenarios show great variation in the costs, timing, and scale of mitigation. All other factors being equal, a lower stabilization target leads to higher mitigation costs and calls for an earlier start with mitigation measures and a greater reduction in greenhouse gas emissions. The UN Framework Convention on Climate Change (FCCC) calls for 'stabilization of greenhouse gas concentrations in the atmosphere at a "safe" level'. It is not clear what a 'safe' level is. The CO_2 concentration threshold of 550 ppmv by the end of the century has often been regarded as a target for stabilization and in the post-SRES exercise the 550 ppmv stabilization target also appeared to be the most popular among modellers. However, there is evidence of the need to be more conservative and consider a safe level to be a concentration of CO_2 around 450 ppmv (Berk et al., 2001; see also Metz et al., Chapter 13, below); see also the discussion in Sections 4 and 5, above. This roughly corresponds to a doubling of the pre-industrial level and is still 50 per cent above current levels. It should be borne in mind that, although most stabilization scenarios focus on CO_2 alone, it is the combined effect of *all* greenhouse gases that counts. Radiative forcing can be expressed in terms of CO_2 equivalents. Stabilizing at some level of equivalent CO_2 concentration implies maintaining CO_2 concentrations at a lower level. A CO_2 concentration of 450 ppmv approximately translates into a 550 ppmv concentration of CO_2 equivalents.

Different emission paths may lead to the same future concentration. This draws attention to the timing issue. One has to balance the costs of early and more gradual action against the costs of later, more rapid forced action. Based on 'integrated assessment' models, some authors have argued that the least costly way to achieve concentration stabilization would be to let emissions continue unconstrained for a certain period of time, followed by drastic cuts later (Wigley et al., 1996). However, Goulder and Mathai (1997) reach the conclusion that a strategy of early and modest abatements may prove less costly if learning by doing is important. Timing is not just a matter of costs. The issue is also strongly related to the carbon cycle: how emissions translate into concentrations. The time window narrows with the size of the reduction efforts. Higher emissions in the reference scenario allow less delay in abatement, and more stringent stabilization targets also require earlier action. Given CO_2 stabilization targets of 450 or 550 ppmv and assuming no ups and downs in concentrations during this century there is little scope for delaying action. In the post-SRES work all 450 ppmv stabilization scenarios require a reduction of at least 20 per cent of global energy-related CO_2 baseline emissions before 2030. A target of 350 ppmv is not feasible even if cuts in emissions are immediate and drastic.

As an illustration, Figure 3.3 shows emissions for two characteristic scenarios. The solid lines represent baseline emissions. The dotted lines denote global emissions leading to a CO_2 concentration of 450 ppmv at the end of the century. The dashed lines are the baseline emissions of developing (non-Annex B) countries. Figure 3.3A refers to the 'dirty' A2 scenario, Figure 3.3B to the 'clean' B1 scenario. Both scenarios show that significant departures from baseline emissions are needed to reach stabilization around 450 ppmv. In the A2 scenario, emissions at the end of the century need to be less than one-tenth of the baseline emissions. Less dramatic reductions are needed in the B1 case, but even in this 'clean' scenario emissions in 2100 are well below the 2000 level. The figure suggests that achieving stabilization at 450 ppmv will require emission reductions in Annex B countries that go significantly beyond the Kyoto Protocol commitments.

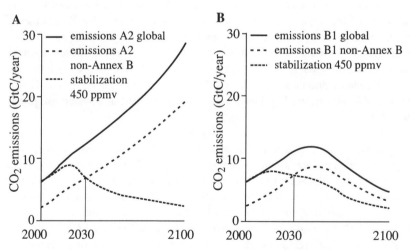

Figure 3.3 CO_2 emissions in A2 and B1, 2000–2100 (reference emissions and 450 ppmv stabilization profile)

From the figure above it is clear that, in this specific case, there is a need for emissions to diverge from baseline levels in developing countries. In both 'worlds', emissions by non-Annex B in the reference case exceed global emissions in the stabilization path. More generally, the stabilization target and the baseline emission level both determine when developing countries' emissions have to diverge from their baseline emissions. The lower the stabilization target, the earlier developed countries would need to reduce greenhouse gas emissions: except in some of the low emissions scenarios (A1T and B1), there is a need for developing country emissions to diverge from baseline levels during the 21st century for even 650 or 750 ppmv stabilization. Most of the post-SRES modelling analysis suggests that developing countries need to prepare to curb baseline emissions in the first half of the 21st century. Even if, strictly speaking, reduction in industrialized countries (Annex B) alone would be sufficient for stabilization of greenhouse gas concentrations, there is still sufficient reason to induce developing countries (non-Annex B) to join a global agreement. Abatement options in developing countries are much cheaper than in the industrialized world. For efficiency reasons alone, it pays to have part of the reductions take place in non-Annex B countries.

The need for participation by developing countries draws attention to the issue of burden sharing (see also Berk et al., Chapter 10, below). The scenarios strongly suggest the need to consider the allocation of emission reductions between North and South. Maximum use of flexible mechanisms seems to be the obvious means to restrict global costs (see also Woerdman et al., Chapter 7, below). For instance, a regime of global permit trading allows emissions to be reduced where it is cheapest to do so. Crucial for the regional distribution of costs is the distribution of property rights to emissions, the way assigned amounts are distributed. For the total costs, according to Coase (1960), the allocation of assigned amounts is irrelevant in a system of global permit trading.

A frequently advocated rule of equity for burden sharing is based on equal per capita emission rights. In this case, there is a global emission target and regional amounts are assigned according to population. Figure 3.4 shows emissions and allocation of per capita emission rights in the 'dirty' A2 and the 'clean' B1 scenario respectively. Both baseline emissions and emissions leading to a stabilization at 450 ppmv by the end of the century are given for the year2050. In addition, assigned amounts based on an equal per capita rule are shown. In both scenarios, per capita emissions in industrialized countries (Annex B) remain much higher than in developing countries (non-Annex B). To achieve stabilization, emissions have to fall in the North and the South. Developing countries reduce even more in relative terms. This is due to the cheap options in those regions. Emissions in developing countries stay well below assigned amounts. The South sells emission rights to the North. In other words, they get paid to reduce more than their emission targets. In A2, the associated costs, in terms of income loss compared to the baseline level, are almost 5 per cent for industrialized countries and 3 per cent for developing countries. In B1 income losses for both aggregate regions are much smaller (0.5 per cent).

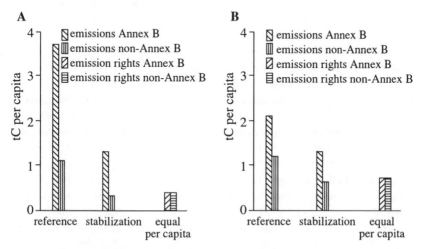

Figure 3.4   Per capita CO_2 emissions and assigned amounts in A2 and B1, 2050 (reference emissions and 450 ppmv emissions and assigned amounts)

Besides this benchmark equity rule, other allocations might be considered (see Rose et al., 1998). For example, emission rights could be based on historic emissions (grandfathering) or emissions could be related to income levels. Additional emission rights or assigned amounts can also be used as a means to soften the negative effects of climate policy. Even with an equal per capita distribution of emission rights, developing countries will suffer an income loss. Industrialized countries could grant more rights to the South. The benefits from selling extra emission permits could make up for the economic loss caused by emission reduction. Manders and Tang (2001) show that the cost for industrialized countries would on average double if developing countries have to be fully compensated.

In general, costs of stabilization, measured as a loss of income compared to the baseline level, are in the order of full percentages. Two qualifications are relevant here. Costs for a specific region or sector may be much higher.

For example, oil-exporting countries or energy-intensive industries may face high costs. We have also ignored adjustment costs, which may be quite high (OECD, 1999).

It is clear that the choice of the future emission baseline is critically important in determining the potential for, and cost of, different mitigation policies. The set of feasible mitigation instruments and the timing, regional allocation and costs of greenhouse gas reductions strongly depend on the qualitative and quantitative characteristics of the assumed baseline. For instance, the greater wealth and turnover of capital stock implied in high economic growth baselines allow for greater potential improvements in energy intensity in the corresponding mitigation scenarios. The different SRES baseline worlds require different measures to stabilize at the same level. The 'high emission' worlds (A1F1, A1B and A2) require a wider range of measures, which are more strongly implemented, than 'low-emission' worlds (A1T, B1 and B2). For example, energy-efficiency improvements in all sectors, the introduction of low-carbon energy and afforestation would all be required in the A1F1, A1B and A2 worlds in the first half of the 21st century. Additional introduction of advanced technologies in renewable energy and other energy sources have to play a role in the second half of the century. Specific properties of 'storylines' associated with particular baselines may restrict available measures and policies, limit regional participation and make early reduction unlikely (for example most defining features of the SRES' 'regionalized' A2 world would complicate the early development of effective mitigation policies).

No single measure will be sufficient for the timely development, adoption and diffusion of mitigation options to stabilize atmospheric greenhouse gases. Rather, a portfolio based on technological change, economic incentives, and institutional frameworks might be adopted. Large and continuous energy-efficiency improvements and afforestation are common features of mitigation scenarios in all the different SRES worlds. Introduction of low-carbon energy is also a common feature of all scenarios, especially biomass energy over the next one hundred years and natural gas in the first half of the 21st century. Reductions in the carbon intensity of energy have a greater mitigation potential than reductions in the energy intensity in the latter half of the 21st century, while the opposite applies in the beginning of the century. In an A1B or A2 world, either nuclear or carbon sequestration would become increasingly important for greenhouse gas concentration stabilization, the more so if stabilization targets are lower. Solar energy could play an important role in climatic stabilization in the latter half of the 21st century, especially for higher emission baseline or lower stabilization levels.

Robust policy/technological options include technological efficiency improvements for energy supply and use, social efficiency improvements, renewable energy incentives and the introduction of energy price incentives such as a carbon tax. Energy conservation and reforestation are reasonable first steps, but innovative supply-side technologies will eventually be required to achieve stabilization of atmospheric CO_2 concentration. Possibilities include using natural gas and combined-cycle technology to bridge the transition to more advanced fossil and zero-carbon technologies such as hydrogen fuel cells. However, even with emissions control, some modellers found that energy systems would still be dependent on fossil fuels over the next century.

7   CONCLUDING REMARKS

Scenarios are an important means of investigating appropriate policy interventions, particularly if one is interested in global, long-term and complex issues such as climate change. Recently developed scenarios show that the range of possible futures is indeed very wide; accordingly, the expected climatic changes differ dramatically between scenarios, as does vulnerability to climate change. The effort needed to curb greenhouse gas emissions thus differs greatly from one scenario to the next. There are various reasons for wanting to reduce emissions. On economic grounds, there is little reason for ambitious emissions reduction. However, from ethical, ecological or legal perspectives more action is called for.

Stabilization scenarios show great variation in costs, timing and the magnitude of mitigation. Costs of stabilization vary greatly, not only between scenarios but also between types of policy intervention. Developing countries should diverge from baseline emissions, and must participate in an abatement regime if one adopts an ambitious emission reduction target.

No single measure will be sufficient for the timely development, adoption and diffusion of mitigation options to stabilize atmospheric greenhouse gases. Rather, a portfolio based on technological change, economic incentives and institutional frameworks might be adopted. Decision-makers can greatly change the costs of emission reduction by putting different emphases on sectors, countries and technologies. Decision-makers can also change emission reduction costs by pushing developments from one scenario to the next. It may be cheaper to move to a 'clean' world than stabilize in a 'dirty' world.

The bottom-line conclusion is not that the future is ours to make. It is true that mankind will shape its own future, but it is an illusion to assume that people will act in unison. Nor is the bottom-line conclusion that everything is too uncertain and too dependent on developments beyond our control. Rather, the conclusion is that an efficient climate policy, working at all relevant levels (ranging from development trends to macro-technology choice to micro-incentives) and with a portfolio of instruments, can achieve a relatively great deal at relatively little expense.

REFERENCES

Berk, M., J. Minnen and B. Metz (2001), *COOL Global Dialogue, Synthesis Report*, RIVM.

Bollen J., A. Manders and H. Timmer (2000), 'The Benefits and Costs of Waiting: early action versus delayed response in post-SRES stabilization scenarios', *Environmental Economics and Policy Studies (special issue)*, 3(2), 143–58.

Coase, R. (1960), 'The Problem of Social Costs', *Journal of Law and Economics*, **3**, 1–44.

CPB (1992), *Scanning the Future: A Long term Scenario Study of the World Economy, 1990–2015*, The Hague: SDU Publishers.

Goulder, L. and K. Mathai (2000), 'Optimal CO_2 Abatement in the Presence of Induced Technological Change', *Journal of Environmental Economics and Management*, **39**, 1–38.

Hasselmann, K., S. Hasselmann, R. Giering, V. Ocana and H. von Stoch (1997), 'Sensitivity Study of Optimal CO_2 Emission Paths Using a Simplified Structural Integrated Assessment Model (SIAM)', *Climatic Change*, **37**, 345–86.

IPCC (2000), *Emission Scenarios*, N. Nakicenovic et al., Cambridge: Cambridge University Press.

IPCC (2001a), *Climate Change 2001: The Scientific Basis*, Contribution of Working Group I to the Third Assessment Report (TAR), Cambridge: Cambridge University Press.

IPCC (2001b), *Climate Change 2001: Impacts, Adaptation and Vulnerability*, Contribution of Working Group II to the Third Assessment Report (TAR), Cambridge: Cambridge University Press.

IPCC (2001c), *Climate Change 2001: Mitigation*, Contribution of Working Group III to the Third Assessment Report (TAR), Cambridge: Cambridge University Press.

Kemfert, C. and R.S.J. Tol (2001), *Equity, International Trade and Climate Policy*, Research Unit Sustainability and Global Change SGC-5, Centre for Marine and Climate Research, Hamburg University, Hamburg.

Kolstad, C.D. (1998), 'Integrated Assessment Modelling of Climate Change', in W.D. Nordhaus (ed.), *Economics and Policy Issues in Climate Change*, vol. 2, Washington, DC: Resources for the Future, pp. 63–286.

Kriström, B. and P. Riera (1996), 'Is the Income Elasticity of Environmental Improvements Less Than One', *Environmental and Resource Economics*, **7**, 45–55.

Manders, A. and P. Tang (2001), 'Obstacles in the Climate Policy Arena', *CPB Report 2001/2, 44–49*, CPB Netherlands Bureau for Economic Policy Analysis, The Hague.

Manne, A.S. and R.G. Richels (1998), 'On Stabilizing CO_2 Concentrations – Cost-Effective Emission Reduction Strategies', *Environmental Modeling and Assessment*, **2**, 251–65.

Morita, T. (ed.) (2000), 'Long-term scenarios on socioeconomic development and climate policies', *Environmental Economics and Policy Studies (special issue)*, **3**(2).

Morita, T., N. Nakicenovic and J. Robinson (2000), 'Overview of Mitigation Scenarios for Global Climate Stabilization based on New IPCC Emission Scenarios (SRES)', *Environmental Economics and Policy Studies*, **3**(2), 65–88.

Nordhaus, W.D. (1992), 'An Optimal Transition Path for Controlling Greenhouse Gases', *Science*, **258**, 1315–19.

Nordhaus, W.D. (1994), *Managing the Global Commons: The Economics of Climate Change*, Cambridge: The MIT Press.

OECD (1999), *Action Against Climate Change*, Paris.

Peck, S.C. and T.J. Teisberg (1992), 'CETA: A Model for Carbon Emissions Trajectory Assessment', *Energy Journal*, **13**(1), 55–77.

Peck, S.C. and T.J. Teisberg (1994), 'Optimal Carbon Emissions Trajectories When Damages Depend on the Rate or Level of Global Warming', *Climatic Change*, **28**, 289–314.

Rose, A., B. Stevens, J. Edmonds and M. Wise (1998), 'International Equity and Differentiation in Global Warming Policy', *Environmental and Resource Economics*, **12**, 25–51.

Tol, R.S.J. (1999a), 'Safe Policies in an Uncertain Climate: An Application of FUND', *Global Environmental Change*, **9**, 221–32.

Tol, R.S.J. (1999b), 'Spatial and Temporal Efficiency in Climate Change: Applications of FUND', *Environmental and Resource Economics*, **14**(1), 33–49.

Tol, R.S.J. (1999c), 'The Marginal Costs of Greenhouse Gas Emissions', *Energy Journal*, **20**(1), 61–81.

Tol, R.S.J. (1999d), 'Time Discounting and Optimal Control of Climate Change – An Application of *FUND*', *Climatic Change*, **41**(3–4), 351–62.

Tol, R.S.J. (2001), 'Equitable Cost–Benefit Analysis of Climate Change', *Ecological Economics*, **36**(1), 71–85.

Tol, R.S.J. and H. Dowlatabadi (forthcoming), 'Vector-borne Diseases, Climate Change, and Economic Growth', *Integrated Assessment*.

Wigley T., R. Richels and J. Edmonds (1996), 'Economic and Environmental Choices in the Stabilization of Atmospheric CO_2 Concentrations', *Nature*, **379**.

4. Global trends and climate change policies

Peter Nijkamp and Harmen Verbruggen

1 INTRODUCTION

This chapter aims to map out the drivers which are likely, from a global perspective, to exert an anthropocentric impact on climatological conditions in our world. It does so by identifying megatrends with respect to institutional change, internationalization and economic integration, technological progress, developments in transport systems, demographic transformations, cultural change and shifts in international business strategies. The chapter also assesses the relevance of these megatrends for designing and implementing internationally co-ordinated climate change policies.

The idea of sustainable development as a policy and analytical concept is very much 'en vogue' nowadays. It has led to numerous policy initiatives and an avalanche of literature (for an overview, see Van den Bergh, 1996). Sustainable development will not arrive like 'manna from heaven', but requires changes in behaviour, economic mechanisms, institutions and technology (see for example Ayres and Ayres, 1997). In this context, technological change towards environmentally-benign modes is often regarded as a sine qua non (see for example Dosi et al., 1990; Lundvall, 1992). Major impediments to these transformations are often inertia and a lag in adoption mechanisms for environmental technology as a result of lock-in structures in path-dependent systems. This naturally has serious implications for an effective and proactive sustainable policy at all levels from local to global. All this is especially cogent for climate change. Despite the awareness of the potential severity of the climate change problem, its 'remote' nature often provokes insufficient policy action and public support at local or national levels. Local quality of life often comes higher in the hierarchy of needs than climate change.

In general, the transboundary character of many pollutants creates the need for an international regulatory system for spatial externalities. The emergence of international environmental agreements is closely linked to the recognition of a collective problem which cannot be solved by means of segmented or individualized national or local policies, or even by bilateral co-operation. The traditional economic wisdom on how to cope with environmental externalities fails in the case of transboundary pollutants because of differences in regulatory and legal systems, problems with identifying and addressing polluters (especially when there are many of them), high transaction costs in implementing policies and differences in the economic interests of the countries or stakeholders involved. The complex nature of global environmental change therefore calls for effective policy mechanisms of a transnational nature, as is evidenced by the current climate change initiatives – and frustrating results – following the Kyoto Protocol.

A change towards more sustainable modes of production, consumption and lifestyle, which are also in harmony with the objectives of climate change policy, would require various institutional, technological and behavioural adjustments or innovations. But such changes cannot be implemented at the same time in all regions of the world. As a consequence, we would face a first-comers versus late-comers dilemma (see also Castells and Nijkamp, 2001), a situation which previously emerged in Europe after the adoption of the agreement on Long-Range Transboundary Air Pollution in Europe. Now that we seem to have sufficient scientific evidence to justify and support strict climate change policies, the question is what insights are needed in the areas of economics and social sciences to support the international climate policy-making process. Given these observations, this chapter addresses just a small number of aspects of the required insights, namely those related to the interface between more or less autonomous global trends and the necessary changes in climate change policies. The focus will in particular be on those megatrends which may be directly or indirectly conducive to or thwart attempts to combat climate change. We argue that an institutional redesign of the governance structures of climate change may be needed to ensure long-term sustainable development.

2   THE HUMAN ACTIVITY SPACE IN A GLOBAL PERSPECTIVE

Ever since the world began mankind has seen its task as being – or has been forced – to expand its action radius in search of new development prospects. The present age of globalization is essentially the latest stage in the long history of civilization, which will be marked by intensive (physical and virtual) interactions on a worldwide scale. Globalization has become a fashionable term in policy discussions and in the scientific literature, used as a concept that refers to structural transformations in our economy through which economic activities are not only connected to their local base but are linked at a global level by means of various ramifications of a political, socio-cultural and economic nature. This interface between patterns of local activity and global developments is a complex phenomenon which has only recently become the subject of scientific investigation. Gaps in knowledge are thus inevitably abundant and much research would be needed to offer a satisfactory and complete picture. Even scarcer is insight into the intricate relationship between changes at the global level and their implications for the environment, not to mention changes in the climate on our earth. There are evidently multiple interwoven forces at work (such as internationalization of markets, accelerated development and diffusion of technologies, worldwide communication and mobility, uniformity in consumption patterns and lifestyles, alliances of globally-operating firms and 'nomadic' behaviour by firms on a worldwide scale), which may act as both causes and consequences of globalization at various geographical levels and on different time-scales (cf. Van Veen-Groot and Nijkamp 1999, 2000).

An important structuring factor from a global perspective is the change in the institutional mechanisms shaping our world economy, such as the World Trade Organization (WTO). Despite much criticism, each country wants to be a stakeholder in such organizations for the sake of prestige and the economic benefits. The same holds for the IMF, the OECD and the World Bank. But these organizations have not played a dominant role in environmental or climate policies. The UN has become more or less the advocate of sustainable development on a global scale, but has manifested insufficient power to enforce strict legislation, as has been witnessed at many recent climate conferences. Hence negotiations take a long time and often lead to disappointing agreements with a low profile. Short-run interests seem to play a more important role than long-term benefits for society at large. The pace of development and implementation of international environmental and climate policy is therefore slow due to a lack of political will (or courage) and social acceptance. As long as there is no international 'carrot- and-stick' system for enforcement of adherence to globally-accepted environmental standards, it will be hard to develop successful strategies for climate change. International environmental and climatic governance is thus a major challenge from an institutional perspective.

Clearly, a multi-faceted phenomenon like globalization will have a multiplicity of environmental and climatological implications. Although many aspects of these implications have been studied in recent years, it was often in a fragmented manner. An OECD (1997) study distinguished four different effects:

- scale effects: globalization will lead to larger world output;
- structural effects: globalization will generate shifts in the composition and location of production and consumption activities;
- technology effects: a competitive world will promote different technology paths;
- product costs: in a growing world economy different product mixes will be produced and consumed, reflecting regional differences in product combinations in relation to different adoption patterns of new technologies.

It seems plausible that the effects – individually and in combination – will have important consequences for our physical world, in both the short term and in the long run. Such effects may be devastating for sustainable development, but may also create new opportunities for environmentally-benign modes of living, working and producing (as can be seen in the environmental Kuznets curve; see De Bruijn and Opschoor, 1997). It should be noted that the implications of these four types of effects for the emission of greenhouse gases in the long term are not at all clear, given the overwhelming uncertainty about the impact of new environmental technologies and technological advances in general. Furthermore, even if it were possible to estimate the extent to which global warming will take place, its effect on the stability of ecosystems and on the climatological system is hard to assess. What we nowadays observe in research and policy practice is a reliance on scenario analysis, which might give the false impression that greater certainty is being created. But in a way scenario analysis takes uncertainty almost to its extremes, since a scenario is only a feasible – but certainly not by definition a plausible or probable – future image. A scenario is essentially part of a learning model, which acts as a sort of flight simulator for policymakers training them to cope with uncertainty, rather than an empirically-tested assessment tool that can be used for future predictions.

The overwhelming use of scenarios in climate change analysis is thus a clear reflection of the absence of sound knowledge of future empirical states of our world (for example in terms of technology, international co-operation.). It therefore makes sense to offer a qualitative, systematic overview of global megatrends which will most likely have an impact on the volume of greenhouse gases and hence on our climatological system, without knowing whether such changes are good or bad for society or the environment. Such an attempt will be made in the next section. We will then try to systematically assess the implications for the design and implementation of an international climate policy by offering a list of qualitative drivers.

3   MEGATRENDS IN A GLOBAL VILLAGE

3.1   Introduction

The past decade has witnessed a vivid debate on various questions relating to climate change. For example, is the climate changing (for example what are the relevant indicators and on what time-scale?). If it occurs, will climate change manifest itself equally in all regions of our world (for example is there an equity issue involved in the effects of climate change?). If it exists, will climate change follow its own natural long-term cycle or rhythm (for example is human behaviour a factor in climate change, and if so, to what extent?). Last but not least, is climate change good or bad (for example, would it have unforeseen benefits, in temperate or cold regions for instance, in addition to costs?) A consensus is gradually emerging among climatologists that the role of the emission and concentration of greenhouse gases is not neutral in climate change, irrespective of whether those greenhouse gases can be absorbed by natural long-term adjustment mechanisms in our climate system. There is complete agreement that a rise in emission rates of greenhouse gases such as CO_2 is a direct effect of human activity. It is therefore important to identify classes of driving forces of an anthropocentric nature which have an impact from a global perspective on the production, emission, distribution and concentration of such greenhouse gases. It should be recognized that the sources of the man-made contribution to climate change are to be found in the behaviour of some six billion inhabitants of our earth, all of whom have totally different lifestyles and consumption and mobility patterns and whose geographic distribution is unequal. It would require a massive research effort to identify the drivers of micro-behaviour of so many people. However, the earth's ecosystem is an interwoven system at various levels, so that local activities ultimately have an influence – directly or indirectly – on the global environmental system. The same applies to local economic activities in a particular region, which are – directly or indirectly – influenced by or have an influence on the world economic system. Against the background of this concept of a 'global village', it seems fair to state that worldwide human activities are not neutral regarding man-made factors in our climate system. This calls for the collection of up-to-date databases on human activities and environmental implications on a spatially differentiated scale (for empirical data, see for example Stanners and Bourdieu, 1995).

From an international perspective, it is possible to distinguish megatrends which are likely to act as drivers for the occurrence of global warming. The list is almost endless, and depends on the disciplinary angle taken, which may range from ethics to astronomy. In the context of the present chapter, we have selected eight driving forces which are directly linked with human behavioural patterns. The megatrends that we discuss here are:

1. institutional change;
2. internationalization and economic integration;
3. rapid technological progress;
4. the emergence of a knowledge economy;
5. improvement of logistic and transport systems;
6. demographic developments and transformations;
7. cultural shifts;
8. international business strategies.

3.2   Institutional Change

Particularly since the end of the Cold War, governments around the world have resorted to market forces to improve the functioning of their economies and the efficiency and quality of government services and utilities. The facilitating processes of liberalization and privatization involve radical institutional changes at global, regional, national and local levels. Older, more rigid institutions and structures are being replaced by new ones to create new markets or improve the functioning of existing ones. The greatest institutional change can be witnessed at a national level, the former centrally-planned economies being the prime example. Regional organizations, such as the European Union (EU) and the North American Free Trade Agreement (NAFTA) have also moved with the times. Institutional change in the international organizations that have been established since the Second World War to facilitate and govern global economic integration is clearly proceeding at a slower place. The IMF and the World Bank have always been fervent promoters of market forces, but a further liberalization of world trade through the WTO seems to have been blocked by political opposition from various directions, while the Multilateral Agreement on Investments, negotiated under the auspices of the OECD, has been terminated. As already mentioned, however, it has to be acknowledged that the process of building global environmental institutions is proceeding even slower. In addition, national environmental policies and the design of international environmental policies increasingly interfere with existing international regulations and agreements, such as those relating to international trade, investment and economic integration. Institutional renewal at the global level encounters serious economic and political obstacles, mainly due to North–South contrasts and political tensions in the North between the USA and the EU.

The undesirable result is that international economic relations can develop undisturbed, without adequate backing and regulation by an international system of environmental governance. Corporate climate governance will take a long time to develop. Emerging institutions and regulations will evolve as a response to the severity of the problem and worldwide awareness of the climate issues.

3.3 International Economic Integration

Our world is moving towards a complex network economy, in which economic interaction at a global level is gaining in importance. Liberalization of international trade – under the influence of the WTO – is but one example; we see similar developments on a regional scale in the EU, ASEAN, NAFTA, MERCUSOR and Australasia. As trade barriers vanish there is more scope for international trade and foreign investments, which will undoubtedly increase welfare in all the participating countries. For quite some time now world trade has, on average, been rising even faster than the domestic production of many countries. The upward trend in international trade relates not only to goods or services but also to foreign investments (since the 1990s). Competition (product and price competition) is obviously fierce in many domestic markets nowadays. Consequently, the production–consumption–waste chain is not only tending to become longer, but also more international. For example, the average carton of yoghurt – which could in principle be produced by the local economy – needs the trade involvement of some six countries.

From an economic perspective, international trade and product specialization can be ascribed to comparative costs and natural resource availability, although this explanatory model has to be supplemented with elements from the theory of product specialization and monopolistic competition (Dixit and Stiglitz, 1977), technological regimes (Dosi et al., 1988) and agglomeration economies (Fujita et al., 1999). This trend of international economic integration is facilitated by changes in industrial economics which point to the importance of the global 'components' industry (Lagendijk, 1995). Investment flows have followed the worldwide mobility drift; they are not substitutes for international trade, but in fact tend to promote the increase in international flows of goods. International capital markets are nowadays developing at a rapid pace and have a decisive influence on the fate of national economies. International monetary integration, particularly in the EU, is another facilitating factor in this respect, as it contributes to the rise of integrated goods and financial markets.

In conclusion, internationalization of business life is a major trend in a modern network economy. The decline in domestic protection and the access to international markets have induced worldwide growth in trade and investments, accompanied by the emergence of the component industry through which goods have become part of a transnational production chain.

The implications of an open world economy for environmental quality do not seem to be favourable (see for example Copeland, 2000). Clearly, for an economist this is a problematic issue, since the theory of trade teaches us that there are gains to be made from trade as a result of spatial specialization. But the resulting trade may, overall, be detrimental to the environment. In any case, increased trade-related transport generates more pollution. The recent discussion on ecological footprints has highlighted many anomalies in our belief in the benefits of the exchange of goods (see Van den Bergh and Verbruggen, 1999).

3.4 Rapid Technological Progress

Since the 1980s, technological change has played a more important role in increasing the competitive advantage of nations and regions. Technological progress has become a vehicle for accelerated growth of productivity and has laid the foundations for a worldwide increase in welfare. The economic and monetary integration and the challenges of intensified global competition have clearly increased the pace of innovation and diffusion of new products and services. A notable driver in this process has been the emergence of increasing returns to scale, which may facilitate a higher growth path, apart naturally from the cyclical business patterns inherent to market economies. For economic and strategic reasons, research and development efforts are increasingly concentrated in big firms operating internationally. The industrial organization features a mix of outsourcing, mergers, take-overs and strategic alliances. The most dynamic technological clusters include ICT, biotechnology, materials technologies, micro-mechanics and nano technology. The ICT sector has proved to have its own indigenous growth pace and welfare-enhancing impacts, but the technology has also had a pervasive effect through its benefits for other technologies (for example aeronautics).

The production and diffusion of new technologies is clearly not location-independent. There appear to be locations which offer excellent seedbed conditions for new technologies, such as Silicon Valley. Although some technologies are in a strict sense footloose, the local business culture and sense of entrepreneurship nevertheless leads to site-specific incubator patterns of technological change. This all has important consequences at both local and global levels. Technology has become an international good (witness the 'human genomics' project). Often, the main question concerns the conditions under which technology will be rapidly adopted and which countries or regions are the forerunners.

An important question in the era of ICT is whether these technologies will act as a substitute for physical movement. If so, ICT could be a *deus ex machina* for climate policy. But despite several – often popular – claims for the potential of ICT, it has to be noted that the empirical facts do not provide convincing evidence. There is indeed some substitution, but this is usually more than offset by the generation of new movements (for the facts, see for example Salomon and Mokhtarian, 1997).

The knowledge economy, on the other hand, may act as a facilitator of environmentally-benign behaviour and lifestyles. More information may increase awareness of the threats of climate change and speed up the search for sustainable technologies. This Kuznets-based argument could imply that the knowledge economy is favourable for environmental quality in the long run.

3.5 The Emergence of the Knowledge Economy

The service economy has long been considered the final stage in the pattern of development of economies. In many developed economies services now account for more than two-thirds of GDP. The new wave of technological change alluded to above provides a new impulse for the service economy. The knowledge revolution makes everything more precise, smaller, faster and cheaper. Process and product innovations are introduced in rapid succession, and new ICT, telecom and media services are marketed. The impact of this revolution is perceived in terms of the new economy that will free us from inflation and recession. Above all, it is said that the new economy will be virtual (that is immaterial) with environmental degradation becoming a thing of the past. However, whether we will really move in the direction of a knowledge-intensive services economy in which our needs are largely satisfied by intangible services depends greatly on the extent to which activities with a large impact on the environment are replaced by cleaner alternatives. As things stand now, the virtual products of the new economy are not that intangible, generate additional physical flows and are complementary to, rather than substitutes for, the existing product mixes that are produced and consumed. The knowledge economy may even lead to the paradoxical situation that advanced knowledge and rising incomes may generate an unprecedented rise in the demand for goods, thus exerting a devastating impact on the global environment.

3.6   Improvement of Logistic and Transport Systems

At a worldwide level we are observing an enormous improvement in logistics and transportation systems. The design of terminal facilities (for example for global container transport) and the development of hub-and-spokes systems have generated a huge increase in transportation productivity. This trend is of course backed by the growth in international trade as a result of globalization forces. The efficiency increase in the transportation sector and the fierce competition in the sector have led to price wars, which encourage long-distance haulage. The leisure market for air passengers is a good illustration of this trend. Consequently, it seems likely that as a result of various mutually reinforcing trends the volume of transport will show a continual rise in the future. In as far as the transportation technology is unable to develop more environmentally-benign options, one may expect a negative impact on the global environmental quality. It should be noted that, in general, an increase in the efficiency of transport systems may be environmentally-friendly and hence alleviate the severity of global environmental change. The main question is of course whether this relative improvement could be overshadowed in the long run by increases in the volume of physical movements. Various studies suggest that it is plausible that the environmental strain caused by the transport sector will continue to grow (see Nijkamp et al., 1999).

3.7   Demographic Developments and Transformations

In past decades demographic developments have shown a steady increase. It will take at least 50 more years before the world population can be expected to stabilize. Clearly, there will be a major shift in the distribution of the world's population. Without massive migration the now developed countries will inhabit a minority of the world's population in 100 years from now. It is clear that more people will impose a greater strain on environmental sustainability in developing countries. If the technological progress in environmentally-benign products and services lags behind in relation to the growth rate of the population or their material consumption, then a major threat to the global environment can be expected.

Apart from the quantitative increase in the population, we will also observe a qualitative shift. There will be an ongoing shift towards urban modes of living, which will mean the emergence of many mega-cities with more than 10 million inhabitants. As a consequence, the quality of the local environment, including key resources such as water and food, will increasingly become a major source of concern. On the other hand, cities offer great opportunities – because of their scale and agglomeration economies – for increasing efficiency in the use of environmental and energy technology. In conclusion, demographic change in association with settlement change will be a major cause of environmental change on our planet.

3.8 Cultural Shifts

The world economy is not only moving towards an integrated network constellation, but also towards a global culture and a new type of civil society. The openness of cultures has prompted a dissemination of views, perceptions, insights, knowledge and cultural identities, a phenomenon also instigated by the modern ICT sector. Cultural globalization, which does not mean a uniform global culture but rather a blend of global commonalities and local specificities, seems likely to become a major new development. This has also led to the emergence of 'glocalization', a mix of globalization and localization.

The openness of our network society also leads to volatility in attitudes towards and perceptions of culture. Dynamic movements of cultures have become a worldwide feature. As a result, we are also observing a trend towards rapidly changing consumption patterns with a certain degree of standardization (for example the hamburger culture). Fashions happen in quick succession with a permanent demand for new products and designs, leading to a rapid depreciation of the existing stock of consumer goods.

The personal responsibility in an individualized society leads to responsive citizens with a sense of the civil society. Consequently, decentralization and privatization have become the features of a new type of democratic culture in which loyalty and solidarity are observed mainly from the perspective of individual welfare and local interests. Self-governance is becoming a new characteristic feature of this civil society. At the same time, however, the common nature of many resources (for example climate, water) also needs a new form of international governance.

The tension between individual micro-interests and collective macro-interests also creates tensions between global standards and cultural identity at a local level. This has important consequences for the pursuit of sustainable development all over the world. Conflicts over the cross-boundary management of water resources, deforestation and fishing grounds, for instance, pose a severe threat to human security and biodiversity.

3.9 International Business Strategies

Internationally operating private companies play a crucial role in climate change as major emitters of greenhouse gases, as economic powers that cannot easily be neglected as well as for their technological capabilities and capacities to invest in mitigation technologies. The international business community follows the climate policy negotiations with suspicion and tries to protect its interests in a climate policy-induced changing world economy. Until recently, the views of the business community were dominated by the potential losers in the event of decisive international action on climate change, led by fossil-fuel interest groups and organizations representing relatively energy-intensive industries. However, positions are slowly but steadily changing (see Oberthür and Ott, 1999; Van der Woerd et al., 2000). Business NGOs like the World Business Council for Sustainable Development now follow a more active strategy by not opposing climate policies but calling for prudent action that does not disrupt international business too seriously. For example, almost all major chemical companies monitor their greenhouse gas emissions and a growing number have set quantitative targets for the reduction of these gases, either in terms of absolute or relative emission targets. To take another example, big oil companies such as Shell and BP have adopted a more proactive stance towards climate change. They have supplemented their core business of extraction, refining and distribution of oil (products) with the development and exploitation of low-emission and renewable energy sources. Both companies also introduced in-company schemes for CO_2 emission trading. The automotive industry worldwide has also come to see that they must be better prepared for a world of low-emission vehicles and transport modes. Even commercial banks and insurance companies, European-domiciled financial institutions in particular, are showing a growing interest in climate change. Insurance companies have to address financial risks related to climate change and communicate on them. But the banking sector is also exploring proactive business strategies by setting internal energy targets, by developing new services, such as green mortgages for energy-efficient buildings and green investment funds, and by exploring a role for the banks if the Kyoto flexibility mechanisms become fully operational. Leaders in proactive business strategies are the potential winners with new combinations of economic activities. For instance, industries represented by the Business Council for Sustainable Energy constitute a broad technological-industrial complex, comprising research groups, engineering and consultancy firms and contractors active in the development and marketing of energy-efficient appliances and renewable energy technologies.

4 MEGATRENDS AND THE NEED FOR CONDUCIVE CLIMATE CHANGE POLICIES

At first glance, the megatrends identified above portray a global economy where Western patterns of energy-intensive production, modes of consumption, lifestyles and transport rapidly spread across countries. The risk is even larger if the global economy locks itself into this energy-intensive trajectory, energy demand and greenhouse gas emissions will only increase in the decades to come.

However, it is becoming equally clear that the same megatrends contain the seeds of change. It is undeniable that a shift is occurring in economic structures away from manufacturing towards services. This shift will gradually emerge in developing countries and Eastern Europe as economic development proceeds. The service economy is transforming into a knowledge economy and information society. ICT is democratic and there are indications that the knowledge economy will spread worldwide at a much faster pace than any previous technological trajectory. There is considerable technological potential for major productivity increases in food production. Breakthroughs in energy technology are not far away. These major transformations are supported by the promise that rising global welfare will lead to greater concern for the environment and the interests of future generations. This concern is already voiced by the emerging civil society and reflected in proactive business strategies. Indeed, the potential to achieve change and embark on a worldwide sustainable growth path is impressive.

But this potential still has to be realized. As things stand, by far the majority of the new technologies, new environment-efficient products and new ICT-related services either generate additional physical flows or come on top of existing goods and services. The bounds of possibility are increasing and needs are adrift. It has to be realized that technological potential will only lead to a reduction of greenhouse gas emissions if it substitutes for the energy-intensive trajectory in such a way as to outweigh the growth of world population and per capita income levels. This requires a profound transition, whereas the offspring of the new economy are more complementary to existing patterns of production and the ways in which people work and live.

The challenge is to create a worldwide institutional infrastructure and appropriate incentive structures to drive change rapidly and broadly enough for the potential and promise to materialize. The lack of properly functioning institutions is one of the major impediments for an effective climate change policy. Such a worldwide institutional infrastructure would need to meet various criteria.

In the first place, it would need to be backed by sufficient reliable information; lack of information would lead to arguments and controversial policies. Second, given the complex global institutional ramifications, this type of infrastructure has to foster global co-operation, with sufficient attention for differences in economic welfare and in adjustment mechanisms among different players. Such players would not only be governments or public bodies, but also NGOs and the international private business sector. A major task of a new institutional infrastructure would be to act as a vehicle for change, with a view to the identification of creative and widely supported changes. This would not be based on a blueprint for climate change policy, but on learning principles. And finally, enforcement of change would have to be legitimate in our global democratic society.

The realization of the tasks of a new institutional infrastructure would have to be based on sound economic principles that would offer a proper incentive mechanism. This implies, among other things, incorporation of global environmental costs in our current modes of production and consumption. Furthermore, such a scheme would also have to make it attractive for public, private and semi-private bodies to join such a global 'club', as there may be many positive externalities of both a tangible and intangible nature involved, also in the short run. These potential benefits include the creation of a new competitive strength in energy-efficient technologies and renewable energy sources, improved energy infrastructures and greater security and safety of energy supplies, secondary benefits of improved local air quality, renewables-based electrification of rural areas, symbiosis between sinks and biodiversity, more effective transfer of technology, capacity-building and financial and technological assistance for global environmental policies. The rules governing such a mechanism (the instruments, the flexibility mechanisms) would have to be established in parallel to the formulation of global targets on environmental change. In this respect, the Kyoto and post-Kyoto negotiating process has not been very productive. In Kyoto, emission reduction targets per country (group) were negotiated, while the working and scope of the flexibility mechanisms were left largely undecided. This led to stagnation in the process of implementation. It is inevitable that when reduction targets are fixed and binding and the mechanisms are still under negotiation, countries will try to stretch these mechanisms so as to reduce their national compliance cost. In fact, what we have seen in the post-Kyoto negotiating process is the trading off of flexibility mechanisms (including sinks) against reduction targets. This detracts from the credibility of the Kyoto Protocol and limits the potential benefits from materializing. The coming years must be used to define the flexibility mechanisms in a credible and verifiable way and to put them to work. Practical experience is needed to further improve and elaborate these mechanisms, and to try to foster and integrate the seeds of change contained in the megatrends. Clearly, a combination of research, institutional learning and creativity is essential in developing such a new international framework for governing climate change.

REFERENCES

Ayres, R.H. and L.W. Ayres (1997), *Industrial Ecology. Towards Closing the Material Cycle*, Cheltenham, UK and Lyme, US: Edward Elgar.

Castells, N. and P. Nijkamp (2001), 'Transboundary Environment Problems in the European Union', *Journal of Environmental Economics and Policy*.

Copeland, B.R. (2000), 'Trade and Environment: Policy Linkages', *Environment and Development Economics*, **5**(4), 405–32.

De Bruijn, S.M. and J.B. Opschoor (1997), 'Developments in the Throughput–Income Relationship', *Ecological Economics*, **20**(3), 255–68.

Dixit, A.K. and J.E. Stiglitz (1977), 'Monopolistic Competition and Optimum Product Diversity', *American Economic Review*, **67**(3), 297–308.

Dosi, G., C. Freeman, R. Nelson and L. Soete (eds) (1988), *Technical Change and Economic Theory*, London: Pinter Publishers.

Dosi, G., K. Pavitt and L. Soete (1990), *The Economics of Technical Change and International Trade*, Exeter: Harvester Wheatsheaf.

Fujita, M., P. Krugman and A.J. Venables (1999), *The Spatial Economy*, Cambridge, Mass.: MIT Press.

Lagendijk, A. (1995), *Internationalisation of the Spanish Automobile Industry*, Amsterdam: Thesis Publishers.

Lundvall, B.A. (ed) (1992), *National Systems of Innovation. Towards a Theory of Innovations and Interactive Learning*, London: Pinter Publishers.

Nijkamp, P., S. Rienstra and J. Vleugel (1999), *Transportation Planning and the Future*, New York: John Wiley.

Oberthür, S. and H.E. Ott (2000), *The Kyoto Protocol – International Climate Policy for the 21st Century*, Berlin–Heidelberg–New York: Springer-Verlag.

OECD (1997), *Towards a New Global Age*, Paris: OECD.

OECD (1999), *The Future of the Global Economy: Towards a Long Boom?*, Paris: OECD.

Salomon, I. and P. Mokhtarian (1997), 'Modelling the Desire to Telecommuting', *Transportation Research*, **3**(1), 35–50.

Stanners, D. and P. Bourdieu (1995), *Europe's Environment*, Brussels: European Environmental Agency.

Van den Bergh, J.C.J.M. (1996), *Ecological Economics and Sustainable Development: Theory, Methods and Applications*, Cheltenham, UK and Brookfield, US: Edward Elgar.

Van den Bergh, J.C.J.M. and H. Verbruggen (1999), 'Spatial Sustainability, Trade and Indicators: An Evaluation of the "Ecological Footprint"', *Ecological Economics*, **29**, 61–72.

Van Veen-Groot, D.B. and P. Nijkamp (1999), 'Globalisation, Transport and the Environment', *Ecological Economics*, **31**, 331–46.

Van Veen-Groot, D.B. and P. Nijkamp (2000), 'Globalization, International Transport and the Global Environment: a Dutch Perspective', *International Journal of Sustainable Development*, **3**(4), 1–14.

Van der Woerd, K.F., C.M. de Wit, A. Kolk, D.L. Levy, P. Vellinga and E. Behlyarova (2000), *Diverging Business Strategies Towards Climate Change*, Bilthoven: Dutch National Research Programme on Global Air Pollution and Climate Change, Report No.: 410200052(2000).

5. Climate change policy in changing contexts: globalization, political modernization and legal innovation

Henk Addink, Bas Arts and Arthur Mol[1]

---

1 INTRODUCTION

It is often noted, in various contributions in this volume among others, that climate change policy differs in a number of ways from previous forms of international environmental policy, such as bilateral or multilateral treaties on transboundary rivers or regional policies on the sea. Climate change policy demonstrates the emergence of some new legal aspects of instruments and principles, changing relations between 'hard' and 'soft' law, new economic and political relations between nation-states, new roles for non-state actors in policymaking processes, and so on. There is little doubt these innovations are to some extent caused by the specific nature of this environmental problem. Climate change differs from other environmental problems such as solid waste, water pollution and erosion in that it combines three central characteristics:

- the truly global character of the problem, albeit with an unequal distribution of causes and effects;
- the close relationship with a fundamental characteristic of contemporary economic systems: energy consumption;
- the high degree of uncertainty regarding effects for different regions.

But climate change policy, and the innovations we are witnessing in that field compared to earlier forms of international environmental policy, are also structured by a changing context. For example, general processes of globalization, political modernization and legal innovation have changed, and are still changing the context in which national and international environmental policymaking takes place. This rapidly changing context changes the content and substance of national, regional and international law and policy, and vice versa. Hence, this changing context is both a medium for and outcome of climate change policy: it structures climate change policy, but is also the result of the challenges presented by climate change.

This chapter seeks to find answers to the following key questions:

- to what extent have recent transformations in the social, political and legal context of international policy-making – expressed in the concepts of globalization, political modernization, and legal innovation – shaped the nature of international climate policy?
- to what extent has international climate policy in turn contributed to these transformations?

The format of this chapter is as follows. We start with an analysis of how globalization has changed the context of international environmental policy-making, particularly with respect to climate change (Section 2). In Section 3, we then focus in more detail on the politics of climate change by analysing two phases in modern politics: an early and a late phase. Climate change policy is a typical product of, and has contributed to, the late phase of political modernization of international environmental policymaking. Section 4 looks at legal innovations in three central elements of law: the combination of binding norms and flexible instruments, the combination of soft managerial and (relatively) strong enforcement approaches in compliance mechanisms, and the division of tasks at international, regional and national levels. In the conclusion (Section 5) some remarks will be made on the (probable) effects of all these transformations on climate policy in the near future.

2   GLOBALIZATION AND CLIMATE CHANGE

The idea or concept of globalization (see also Nijkamp and Verbruggen, Chapter 4, above) brings together the major social transformations taking place in the contemporary world economy, in international relations and politics and in global culture. The common denominator of the current changes in these three areas is the growing interconnectedness and interrelatedness of social practices and institutions over time and place. Anthony Giddens (1990, p. 64) defines globalization as 'the intensification of world-wide social relations which link distant localities in such a way that local happenings are shaped by events occurring many miles away and vice versa'. Roland Robertson (1992, p. 8) adds to that definition the perception of living in one world: 'Globalization as a concept refers both to the compression of the world and the intensification of consciousness of the world as a whole'.

Although there is still debate about what globalization means, most scholars refer to the onset of significant changes in economic, political, and cultural spheres in the last quarter of the 20th century as well as the emergence of information and communication technology as major factors in it.

2.1 Globalization and Social Change

What – in a nutshell – are the major social changes brought together under the notion of globalization? Economic, and especially financial, markets are increasingly globally structured, although we should not take 'global' too literally. Economic globalization is restricted to a large degree to the triad of the NAFTA (North American Free Trade Agreement) area, the European Union and Japan and some Asian high-performance economies (see Dicken, 1998, for some data on trade, Foreign Direct Investment, financial flows and the like between these three corners of the triad). Globalization in the field of culture refers to both the homogenization of cultural signs and tokens, norms and values around the world and their heterogenization, as global cultural signs are mixed with specific local circumstances and cultures. This occurs via a number of different mechanisms: migration, the global ICT (Information and Communication Technology) networks and media, and the increasing economic exchanges. McDonaldization (Ritzer, 1993) is the critical evaluation of global culture: it stands for the Westernization of the world through the invasion and increasing domination of Western – or sometimes purely American – culture in the cultural institutions and practices of non-Western societies. But at the same time, eating habits and tastes, religious practices, norms and values and arts flow from non-Western societies into Western culture, giving evidence of at least some mutual dependence and influence. Finally, globalization represents a major transformation in international relations, governance and politics. The system of nation-states, based on the principle of national sovereignty and the prime structuring principle in the international political order for centuries, is being increasingly undermined. States are still important entities in international politics but have to face the growing importance of transnational companies and financial institutions, international organizations, and an active (global) civil society as crucial players in a steadily more complex global political arena. This results in new forms of international and global policymaking, among others in the area of the environment and climate change. This chapter focuses especially on this latter dimension of globalization. Before focusing specifically on transformations in supranational policymaking and legal innovations, we will first elaborate on how the environment is connected to processes of (especially political) globalization.

2.2   Climate Change and Globalization

The environment, and more specifically climate change, is related to the ongoing processes of globalization in at least three different ways. First, environmental interests and considerations have enhanced and structured processes of globalization. Global environmental change, and more specifically climate change, has contributed to the perception of living in one world – largely in line with Robertson's definition – and the need to co-ordinate social practices and institutional developments at a supranational level. In the 1990s, environmental interests and considerations reshaped international political processes and principles, brought about innovations in legal systems, and created new (and also consolidated old) dependencies and power relations. But climate change also plays a significant role in economic and cultural dimensions of globalization, for instance through the formation of global (environmental) norms and values, the internationalization of the environmental movement, the standardization of the (environmental) performance of economic actors (ISO standards being one form of codification), the call for harmonization of national environmental standards, the global transfer of environmental technology, and new mechanisms of economic exchange. In that sense, climate change 'triggers' social transformations at a global level, notwithstanding the often fierce opposition to globalization from those who want to protect the global climate from change.

Second, environmental problems are 'structured' by globalization processes, as are the possible solutions to these problems. This is especially true for those problems that fall under the umbrella of global environmental change, climate change being one of them. Economic globalization processes, especially in their neo-liberal form, often enhance the emission of greenhouse gases due to growth of transport, higher levels of economic activity in production and consumption and a growing demand for energy, among others. At the same time, economic globalization and inter-connectedness and the emergence of increasingly universal norms and values concerning the environment make national environmental abatement strategies increasingly problematic. Both individual nation-states and the system of nation-states are weakened in their ability to cope with the growing demand for sustainability in an era of globalization. This is due not only to the global character of climate change, but even more to the growing interconnectedness of the global economic system and the free movement of capital. Climate change as an environmental problem is especially structured by the imbalance between the high dynamics of neo-liberal globalization in economic practices and the much slower development of supranational and global political and legal environmental institutions, against the background of increasingly universal and globally articulated norms and values concerning the environment.

Finally, the close relation between global environmental change and globalization processes has led to the increasing use of environmental resources in global politics and economics. The debt-for-nature or debt-for-environment exchanges, Joint Implementation, the Clean Development Mechanism and other flexible implementation measures as well as the provisions of the Montreal Protocol are all examples of the use of environmental resources by non-OECD countries to secure economic and technological gains as well as national environmental conservation. The environment is increasingly used by countries either as a political resource to acquire a better position in political negotiations or as a resource to gain technological or economic support in the global economy (Miller, 1995; Mol, 2001). The idea is not completely new: developing countries have long used natural resources such as ores and wood to gain economic resources and an important geopolitical position *vis-à-vis* the more developed countries. What is new, though, is that this now extends to emissions, that substantial flows of money from North to South are not always paralleled by reverse flows of natural resources and that the dependencies between countries are no longer strictly economic. Environmental resources have gained a relatively independent position in global exchanges.

2.3 Europeanization: A Specific Regional Form of Globalization

These processes, dynamics and dependencies can be seen in a more intense form in the European Union. The Europeanization process can be seen as a specific case of 'globalization', as the political, economic and cultural forms of interconnectedness are more intensive and more highly articulated than on a truly global scale. As a result, several authors looking for solutions to environmental problems under conditions of globalization point to the European Union as a prime example, in particular because it seems to manage to combine European economic integration with supranational governance, or economic liberalization with an environmental and social agenda (Beck, 1997; Group of Lisbon, 1995; Held, 1995; Mol, 2001). In addition, environmental considerations have played a crucial role in the European Union at various stages of the European integration process: as a formative factor of supranational politics, to legitimize further Europeanization, as a trigger for economic reallocation via the so-called cohesion funds, and as a (perceived) negative side effect of economic integration and liberalization. This has resulted in ambivalence towards

Europeanization in 'greener' member states (and in Denmark in particular). Many of these dynamics can be identified at the global level, either as ideas or as existing social processes.

3   POLITICAL MODERNIZATION AND CLIMATE CHANGE POLICY

Globalization and Europeanization – as elaborated above – definitely have an impact on the way we conduct politics and design policies today. Due to these processes, the nation-state model has lost its exclusiveness and political authority has leaked upwards and downwards, to regions, private organizations and international organizations. This is *not* to say that state powers have been completely eroded and they have lost their sovereignty, as some post-modernist and neo-liberal thinkers seem to argue (Albrow, 1996; Ohmae, 1995). But it does mean that the role, function and power of the state *vis-à-vis* markets and civil societies have been redefined, either at the domestic or transnational level (Van Kersbergen et al., 1999). There is another reason why the nation-state model has lost its exclusiveness: the decline in the credibility of the old, regulatory state-centric approach of policymaking. Instead, today scientists and policymakers are looking for and experimenting with new governance models. At the national level we witness the emergence of concepts such as network steering, co-production, co-steering, interactive planning, self-steering, economic and communicative instruments and the like (Glasbergen, 1998; Godfroij and Nelissen, 1993; Kickert et al., 1997; Van Tatenhove et al., 2000). They all refer to governance styles and policy instruments which are less top-down and state-centric in nature, thereby allowing more room for stakeholder participation by both governmental and non-governmental players. What we will argue in this chapter is that these innovations and changes in styles and instruments are not only popular at the (sub)national level, but are also emerging with increasing frequency at the international level, and thus shaping international climate change policy (Baker, 2000; Golub, 1998).

3.1   Political Modernization

Some academics call this general process of change 'political modernization' (Jänicke, 1993; Van Tatenhove et al., 2000). To grasp the dynamics of this process, which is also relevant for a more historical analysis of climate change policymaking (see below), it is useful to distinguish two phases: an 'early' phase and a 'late' phase in modern politics (see Table 5.1). The first phase (1950s–1980s) is closely linked to the project of modernity itself. Central elements of this phase are the nation-state model, the regulatory state and the manageable society, elements which are, incidentally, highly interrelated. Although international organizations such as the EU and UN emerged in those decades, politics were mainly shaped within the boundaries of the nation-state model. Moreover, the state was considered to be the supreme regulatory body within its boundaries, in both capitalist and centrally planned societies. At the same time, society was believed to be highly 'manageable' by state regulation.

Table 5.1 Phases in 'modern politics'

Early phase	Late phase
Nation-state model	Multi-level governance
Regulatory state	Multi-actor governance
Manageable society	Governance relativism

The second phase (1980s–present) is closely linked to what some call *post-modernity*, and others *reflexive* modernity (Albrow, 1996; Beck et al., 1994). Political modernization is a reaction to the steering and governance problems associated with early modernity. In the first place, the classical nation-state – in so far as it ever existed – could not bring security and prosperity for all, either in the developed or in the underdeveloped world. Nor could it guarantee that it would do so in the future, given its declining capacities and powers in the face of growing globalization and Europeanization tendencies. More and more, monolithic state governance has been replaced by a system of joint multi-level governance, in which regions, nations and international organizations co-determine political outcomes through complex procedures of participation and decision-making (Baker, 2000; Kohler-Koch and Eising, 1999). Secondly, and related to this, the regulatory state has lost credibility due to the crisis of the Western Keynesian welfare state, the failure of state intervention in many developing countries and the collapse of the socialist state model in Eastern Europe in the late 1980s. Third, top-down policy-making has proved to be less effective, efficient and legitimate than was once claimed. Therefore, governance has not only become multi-layered, but also pluralized (Gordenker and Weis, 1995; Kooiman, 1993). This means that the participation and influence of players from the market and civil society – so-called epistemic communities, NGOs and businesses – has recently increased at all levels of policymaking (Princen and Finger, 1994; Haas, 1993).

Governments' main reasons for bringing in non-governmental players are: to mobilize knowledge for improved policymaking, to increase legitimacy, to secure co-operation for enhancing policy implementation, and to privatize state functions in order to cut budgets (Jacobson, 1984; Mazey and Richardson, 1992; Van Noort et al., 1987). Finally, society has turned out to be less manageable than was believed due to its growing complexity. The effects of policy-making are generally unknown, unintended or unforeseen. As a consequence, governance optimism in the early stages of modern politics has been replaced by governance relativism, or even worse, by governance pessimism.

3.2   Climate Change and Multi-level Governance

The process of political modernization is also reflected in climate change policymaking. In the early 1980s, when climate change was first framed as a policy problem and was put on the international political agenda, most policy proposals were based on the regulatory approach (Bodansky, 1993). The dominant idea was that states would jointly formulate general and legally-binding targets and timetables to reduce greenhouse gas emissions based on information provided by experts, and that these would be included in an international treaty to combat global warming (Hurrell and Kingsbury, 1992). States would subsequently implement them at home. At the same time, alarmed by the scientific data on global warming, environmental movements around the world developed their own alternative programmes, to some extent even more 'early modernist' in nature than those of governments. Some proposed strict targets and strong compliance mechanisms, positions which presuppose fairly strong states with sufficient control over the economy and civil society to reshape production and consumption patterns in accordance with ecological principles (A SEED, 1995; CAN, 1991; Greenpeace, 1991).

Although early policy responses reflect some of the above tendencies (for example the Toronto target of a 20 per cent reduction in carbon dioxide emissions compared to 1989 levels, to be achieved by 2005), climate change policymaking has developed differently. In fact, the setting of targets and timetables with the global level as their starting point has never become a top-down hierarchical road in the climate arena. The policy process has been multi-level in nature right from the start. Individual countries and the EU came up with domestic and EU-wide targets in the early 1990s, while UN members engaged in global negotiations to set a single standard at that time (Mintzer and Leonard, 1994). Subsequently, these initiatives slowly converged, culminating in the FCCC in 1992 and the Kyoto Protocol in 1997.

From the perspective of a single country (for example, the Netherlands) climate change policymaking can equally be analysed in terms of multi-level governance. Although the Netherlands set its own domestic emission reduction targets at the end of the 1980s (3 to 5 per cent reduction of carbon dioxide in 2000 compared to 1990 levels), these targets are outdated now, while new ones are determined by global and European agreements (Kyoto Protocol and 'EU bubble'). According to these, in the period 2008 to 2012 the Netherlands should reduce its greenhouse gas emissions by 6 per cent compared to 1990 levels as a contribution to the common EU target of 8 per cent.

This example also shows that the EU has recently acquired a key position in global environmental governance. Whereas in the early phase of modern politics the member states determined their own foreign environmental policy agenda, with only minor co-ordination through Brussels, nowadays the EU has increasingly become a truly global player and has thus reduced the direct influence of its member states on global governance. In return, the implementation of global agreements is also channelled through the EU more than before (see the 'EU bubble'), although it remains a matter of national implementation as well.

3.3 Climate Change and Multi-actor Governance

Climate change policymaking does not just have a multi-level character. From the start, climate change policy has also been multi-actor in nature, although some Asian and Middle-Eastern countries were reluctant to accept the input of non-governmental players in the climate arena, given their sometimes antagonistic relationships with NGOs at home. Due to the rather open participation processes of the UNCED, non-governmental players have always had (some) access to the climate arena (Mintzer and Leonard, 1994). As a consequence, the role of epistemic communities, NGOs and business was significant (Arts, 1998; Boehmer-Christiansen, 1994; Kolk, 1999). For example, without the IPCC, agenda-setting and policymaking on climate change would definitely have been different, and probably weaker. Environmental NGOs such as Greenpeace, WWF, Friends of the Earth and Climate Action Network were able to influence specific elements of climate change policymaking. The same goes for business, although in different ways and on different issues. Whereas the latter mainly tried to block any progress in policymaking in the early 1990s, for example through the Global Climate Coalition, since then it has developed a more neutral, and sometimes even pro-active, strategy (for an in-depth overview of the role of these non-governmental players in climate-change policymaking: see Arts and Cozijnsen, Chapter 14, below).

3.4   Climate Change and the Renewal of State Governance

During the UNCED process in the early 1990s, there was general optimism about the possibilities for dealing with the problem of climate change worldwide, and the FCCC was an expression of that optimism. However, at the start of the new millennium we face the relative inertia of the international political and economic system. For example, most industrialized countries failed to meet the stabilization target of the FCCC, which was to stabilize greenhouse gas emissions at 1990 levels by the year 2000 (the exceptions being Central and Eastern European countries, which could not reach this target because of economic stagnation and decline following the collapse of communism in 1989). Even worse, in many industrialized countries greenhouse gas emissions today are about 10 per cent above 1990 levels. This 'implementation deficit' has, among other things, inspired policymakers to look for other governance approaches. First, greater involvement of market actors and mechanisms and new policy styles that rely more on voluntary and flexible instruments. Second, ambitions have been moderated, as the FCCC and Kyoto Protocol show. Where the FCCC aimed at a stabilization of emissions at 1990 levels in the year 2000, the Kyoto Protocol aims at an overall 5.2 per cent reduction between 2008 and 2012 (again with 1990 as the baseline). These objectives are weaker than those in the the early Toronto Target (20 per cent reduction in 2005). Generally, this weakening of policy targets is explained by the political unwillingness of countries, mainly those in the JUSSCANNZ group (Japan, USA, Switzerland, Canada, Australia, Norway and New Zealand) to adapt their economies to ecological standards. Although this is part of the explanation, moderation of policy is also an expression of the realization that the 'manageable society' no longer exists. One cannot simply proclaim high standards and govern social change from above. A more democratic and step-by-step approach is probably a more adequate strategy, but it implies moderate targets.

Third, targets and timetables have been differentiated in the climate arena, so that those responsible for climate change and those capable of solving (aspects of) the problem carry the burden. This increases the chances that policies will be implemented. This differentiation is most strongly reflected in distinctions between North and South and within the EU.

Similarly, flexibility might enhance implementation and thus effectiveness. Because the enhanced greenhouse effect is a global risk, from a natural science definition of climate change it makes little difference whether emissions are reduced or removed. This is the basic idea behind the flexible instruments: 'If it doesn't matter where emissions are reduced or removed, please do it where it is the most effective and efficient.' This implies that the responsibility and capability of states and stakeholders for policy implementation on the one hand, and the actual place where this is realized are dissociated. Through Joint Implementation (JI), the Clean Development Mechanism (CDM) and emission trading, states and companies may reduce or remove emissions elsewhere, even abroad. This offers them more flexibility in realizing their goals, which in turn increases the chances of policy effectiveness. At the same time, it may reduce transaction costs, monitoring costs, control and enforcement efforts, and the like (see also Jepma and Bandsma, Chapter 6, below and Woerdman et al., Chapter 7, below).

4   LEGAL INNOVATION AND CLIMATE CHANGE POLICY

The climate change problem induces issue-specific legal innovations in the sense of unique combinations of legal aspects that cannot be compared easily with legal aspects of other environmental problems. On the other hand, these legal developments may also lead to more general innovations in international (including regional) and national administrative and environmental law. This section looks at three legal innovations in climate change law: the strained combination of binding norms and flexible instruments; the consequences of this combination for the compliance regime and the development of a 'legal multi-task division' at international, regional and national level.

4.1   The Strained Combination of Binding Norms and Flexible Instruments

The development of norms in the climate change regulations has been a question of 'hop, skip and jump'. A unique situation exists within the Framework Convention on Climate Change. The Framework Convention concentrates in particular on the institutional and procedural aspects of climate change policy. These aspects must be seen in relation to the ultimate objective as stated in Article 2 and the principles set out in Article 3 of the Convention. But in the beginning the relevant question was: what is the legal significance of the objective and the principles when they do not have the consequences of a legally binding target for industrialized countries (Oberthür and Ott, 1999)? At that moment the Convention contained only a reporting obligation and, for the industrialized countries, the aim of returning to their 1990 level of emissions. In 1992 the Convention only established soft obligations for the parties, including the industrialized countries.

After the acceptance of the Convention, it was concluded during COP I (in 1995 in Berlin) that Articles 4.2a and b were not adequate, which led to the decision to set quantified limitation and reduction objectives within specified time-frames. However, it was also stated that no new obligations for developing countries would be introduced. On the basis of this Berlin Mandate, new talks took place in the Ad Hoc Group on the Berlin Mandate (AGBM). The AGBM discussed not only formal aspects (protocol or amendment, Addink, 1997) but especially the targets for the industrialized countries, and whether there should be differentiation between the various types of greenhouse gases. Even during the Kyoto meeting these targets were uncertain, and right up to the final days in Kyoto this aspect had not been solved. The outcome was the Kyoto Protocol in December 1997. The centrepiece of this Protocol is Article 3: the commitment by industrialized countries to reduce their greenhouse gas emissions by at least 5.2 per cent from 1990 levels in the commitment period 2008 to 2012. They can achieve this using both the traditional non-flexible instruments and flexible mechanisms: Joint Implementation, the Clean Development Mechanism and emissions trading. Both strands – quantitative norms and flexible mechanisms – can to a certain extent be seen as an innovation in environmental law when set out in such broad terms. However, they require an accurate approach.

It was fairly unique for parties to have accepted quantitative binding norms for two reasons. The first reason is that with the Berlin Mandate the parties themselves made clear that the Climate Convention was inadequate for achieving real change in domestic climate change policies. The second reason is that the parties themselves established legally binding, quantitative obligations. It is the first time in history that specific, legally binding obligations were established to reduce those greenhouse gases primarily responsible for global warming (Oberthür and Ott, 1999). Other authors point out that the Kyoto Protocol is a unique document in international environmental law because it includes binding commitments requiring countries to reduce their greenhouse gas emissions through the use of flexible market-based mechanisms (Telesetsky, 1999).

There seems to be an internal tension between the (relatively) strict quantitative norms on the one hand, and the flexible mechanisms on the other. Flexibility in a certain governmental context often means lack of transparency, but it also leads to difficulties in relation to supervision, and uncertainty in relation to the obligations. This makes the Protocol rather risky, and may weaken the impact of the targets. Besides the flexible mechanisms, other concessions have been made that will give the parties additional flexibility when fulfilling their quantitative reduction obligations.

These concessions, made primarily to the United States and including one to the European Union, refer to the commitment period, the basket of gases, the differentiated emission targets, the carbon sinks and the bubbling. In interpreting these elements of the Kyoto Protocol, the literature mentions a number of (potential) loopholes, which will reduce the impact of targets which are formulated as quantitative norms (cf. Rolfe, 1998).

These legally binding obligations in the Kyoto Protocol also seem to ignore (or at least push into the background) the objective and the principles of the Climate Convention. There will be ample discussion of the flexible mechanisms, the quantitative norms as well as the qualitative norms of Articles 2 and 3 of the Convention during the process of interpretation of the different obligations of the Convention and the Protocol.

4.2 Balanced Compliance: A 'Soft Managerial' Approach in Combination with a (Relatively) 'Strong Enforcement' Legal Approach

In the theory on compliance with multilateral environmental agreements, or with national and international regulations in a broader sense, two approaches can be distinguished (OECD, 1998). The first is the managerial approach, which relies primarily on co-operative problem-solving. The second is the enforcement approach, which relies on stronger tactics to deter non-compliance or to coerce non-complying states into compliance.

The impression is that the approach which is chosen often relates to the type of obligations that have to be met. If the obligations can be easily met, the managerial approach is often chosen. However, more and more environmental problems call for a substantial change in state behaviour. As international agreements become more ambitious, more formal and coercive enforcement strategies, which raise the cost of non-compliance, will become necessary on the national and international level (Werksman, 1996). In this situation strengthening non-compliance responses is necessary as the interdependence of parties on each other's compliance increases.

The issues of environmental crime, enforcement and compliance are gaining increasing recognition by experts in national and international law (cf. UNEP, 2000). In the context of the climate change regulations, this will have consequences for the compliance mechanisms, but also for the discussion on the enforcement of multilateral environmental agreements in relation to non-compliance. During the negotiations on the Kyoto Protocol there was already growing interest in a more robust compliance system for the climate regime, including tougher responses to non-compliance (Addink, 1997; Werksman, 1998). One reason for this particular attention to enforcement is the societal impact of the Kyoto Protocol at a national level (Cozijnsen and Addink, 1998a and 1998b).

The following elements can be found in the proposal of the compliance working group of the FCCC (FCCC/SB/2000/CRP.15/Rev.2, 20 November, 2000): principles (proportionality, common but differentiated responsibility, efficiency, due process); as general provision, the establishment of a compliance body with two branches (one dealing with facilitation and one dealing with enforcement); procedures of submission, preliminary examination and the proceedings of compliance and regulations on the outcomes and consequences of non-compliance. All these elements are worked out in detail in the proposal.

Furthermore, the literature already mentioned distinguishes the following elements that could be part of the regime for ensuring compliance with the commitments under Article 3 of the Kyoto Protocol: a strong accounting, reporting and review infrastructure, a number of preventive measures and deterrents to non-compliance (Defining Kyoto Protocol non-compliance procedures and mechanisms, October 1999). The different parties to the Kyoto Protocol have made suggestions for the different elements (Addink, 2000). In developing these elements one can build partly on the existing compliance mechanisms, but there is also a need to develop new enforcement mechanisms. For some suggestions in this context, see Van der Jagt, Chapter 11, below.

To conclude, discussions on compliance focus not only on the soft management approach but also on the legal enforcement approach. This conclusion can still be drawn even in light of the decisions (FCCC/CP/2001/L.7, 24 July 2001) taken at COP 6 bis in Bonn in July, 2001.

4.3   The 'Multi-task Division' at International, Regional and National Level

Three levels can be distinguished in relation to the Climate Change Convention and the Kyoto Protocol: the international level, the regional level and the national level. The international regulations impose obligations on the parties. Furthermore, there are special institutions with special tasks in relation to the implementation of the regulations. However, two aspects of the Kyoto Protocol form an innovative stimulus for the implementation of the Protocol at other levels: the regional level, such as the European Union, and the national level.

One of the concessions made in the Kyoto Protocol was to the European Union in relation to Article 4, since even before adoption of the Kyoto

Protocol the 'European bubble' was an important phenomenon (Ott, 1997). Article 4, which is a unique arrangement, establishes a joint fulfillment agreement, which means that parties in a certain region can enter into an agreement about how to meet their obligations. To participate in such an agreement the parties have to meet certain conditions set out in the Protocol. The failure to meet or to continue meeting these conditions has important consequences. For example, if Annex I parties make an agreement fixing the emission level for each of the parties, the agreement has to be notified to the secretariat of the Convention. The secretariat then informs the other parties. This agreement is for the whole commitment period. An important condition is that in the event of failure to reach the emission level by the parties to the agreement, each party is responsible for its own level of emission (Addink, 1997). This last point is especially important for countries, like the Netherlands, which have a lower reduction norm in the bubble than in the Protocol. Consequently, in the event of failure these countries have to fulfil the more stringent reduction target.

In relation to the bubble situation, but also in general, the respective competence of the European Union and the member states is not entirely clear. The question is when there should be a European Union regime and when there should be a national regime. A recent study states that, in the field of environmental policy and legislation, competencies are generally seen as being shared. This means that both the member states and the European Union can conduct external negotiations in this field. The Kyoto Protocol also introduces new national tasks with respect to climate change, leading to far-reaching obligations for national authorities. Two relevant examples can be distinguished. The first is that regulations have to be drawn up not only at the national level but also at decentralized, provincial and local level (Cozijnsen and Addink, 1998b). The second example is that numerous companies will participate in the implementation of the elements of the climate change regulations (Werksman, 1998). Both lines can be seen as a part of developing international administrative law (Vogel, 1992) and environmental law.

The 'three-level approach' involves a multilateral agreement with strong involvement by regional and national levels in design and implementation. Treaties are no longer only designed and implemented at the international level, but increasingly also at the regional, national and even sub-national level. Globalization processes have not only affected the international level of environmental law, but also have a direct effect on the regional and (sub)national levels. This results in new questions of competencies between the levels, for instance in the European situation as regards the competence of the European Union in relation to the member states.

5   CONCLUSIONS

Climate change policy is both affected by and a cause of changes in global politics, governance and law (although probably more an effect than a cause). In this chapter, we analysed how climate change policy has, to some extent, contributed to significant transformations at different levels: globalization, governance and legal frameworks. At the same time we identified how more general changes have structured the way in which climate change policy is designed and implemented. It is not by accident that at the more substantive level of legal innovations climate change policy is more a *causal* factor, whereas at the level of more general globalization processes, climate change policy is to a significant extent an effect of economic, political and cultural changes.

For the future we can safely assume that these trends of globalization, political modernization and legal innovation will continue. This has implications for climate policymaking. First of all, it offers (enhanced) opportunities for 'doing policymaking differently': from global to local perspectives, with new stakeholders involved, and on the basis of well-balanced legal mechanisms of instruments, norms and compliance. Second, international climate policymaking will become even more complex, as more policy levels, more stakeholders, more interests and more legal aspects become involved. This raises questions of governability. Third, structural processes on a global scale make it ever more difficult – though not impossible – for individual states to act unilaterally. Despite US pressure to blow up the Kyoto Protocol, the rest of the world community decided otherwise in Bonn in 2000; but the US did manage to stay out of the protocol. Finally, the legitimate need for differentiation, flexibility and legal certainty on the one hand, and on the other the legitimate need for short-term policy successes – in terms of real emission reductions – may frustrate progress in international climate policy in the future, as these two needs may clash. Therefore we need policymakers with knowledge and inspiring visions not only on climate issues themselves, but also on the general processes of globalization, political modernization and legal innovation, in order to match these needs.

NOTE

1. The authors wish to thank J.A.E. van der Jagt, the editors and two anonymous referees for their useful comments on an earlier version of this chapter.

REFERENCES

A SEED (1995), *The Greenhouse Gathering*, Berlin: A SEED

Addink, G.H. (1997), *Norms and Enforcement of the Climate Change Convention*, Japan Centre of International and Comparative Environmental Law, UNFCCC Protocol Working Group 1997: Climate Change and the Future of Mankind, Pre-Cop 3 International Symposium on Legal Strategies to Prevent Climate Change, Tokyo.

Addink, G.H. (1999), *Implementation and Enforcement of the Kyoto Protocol After Buenos Aires November 1998*, Jahrbuch des Umwelt- und Technikrechts, Berlin: Erich Schmidt.

Addink, G.H. (2000), 'Joint Working Group Compliance on the Kyoto Protocol: An Overview of Suggestions on Compliance', in *El Protocol de Kyoto*, San José (C.R.): Instituto de Investigaciones Juridicas, Universidad de Costa Rica.

Albrow, M. (1996), *The Global Age. State and Society Beyond Modernity*, Oxford: Polity Press.

Arts, B. (1998), *The Political Influence of Global NGOs. Case Studies on the Climate and Biodiversity Conventions*, Utrecht: International Books.

Arts, B. (2000), 'New Arrangements in Climate Policy', *Change*, **52**, pp. 1–3.

Baker, S. (2000), *Environmental Governance in the EU,* Quebec: IPSA Paper.

Beck, U. (1997), *Was ist Globalisierung? Irrtümer des Globalismus-Antworten auf Globalisierung*, Frankfurt am Main: Suhrkamp

Beck, U., A. Giddens and S. Lash (eds) (1994), *Reflexive Modernization. Politics, Tradition and Aesthetics in the Modern Social Order*, Oxford: Polity Press.

Bodansky, D. (1993), 'The United Nations Framework Convention on Climate Change: A Commentary', *Yale Journal of International Law*, **18**(451), 453–559.

Boehmer-Christiansen, S. (1994), 'Scientific Uncertainty and Power Politics: The Framework Convention on Climate Change and the Role of Scientific Advice', in B. Spector, G. Sjöstedt and I. Zartman (eds), *Negotiating International Regimes. Lessons Learned from the United Nations Conference on Environment and Development (UNCED)*, London: Graham & Trotham/Martinus Nijhoff, pp. 171–80.

CAN (Climate Action Network), *Position Statement, Second INC Meeting*, 20 June 1991.

Cozijnsen, C.J.H. and G.H. Addink (1998a), 'The Kyoto Protocol under the Climate Convention: Commitments and Compliance', *Change*, **43** (August–September), 5–8.

Cozijnsen, C.J.H. and G.H. Addink (1998b), 'Het Kyoto-Protocol onder het Klimaatverdrag: over de inhoud, de uitvoering en de handhaving van afspraken', *Milieu en Recht*, **1998**(6), 152–59.

Dicken, P. (1998), *Global Shift. Transforming the World Economy* (3rd edition), London: Paul Chapman/Sage.

Giddens, A. (1990), *The Consequences of Modernity*, Cambridge: Polity Press.

Glasbergen, P. (ed.) (1998), *Co-operative Environmental Governance: Public-Private Agreements as a Policy Strategy*, Dordrecht: Kluwer Academic Publishers.

Godfroij, A.J.A. and N.J.M. Nelissen (eds) (1993), *Verschuivingen in de besturing van de samenleving*, Bussum: Coutinho.

Golub, J. (ed.) (1998), *New Instruments for Environmental Policy in the EU*, London: Routledge.

Gordenker, L. and T. Weis (1995), 'Puralising Global Governance: Analytical Approaches and Dimensions', *Third World Quarterly*, **16**(3), 557–87.

Greenpeace International (1991), *Preventing A Climate Holocaust: A National Legislative Agenda*, Proposals Submitted by Greenpeace International to the Second Session of the Intergovernmental Negotiating Committee on a Draft Framework Climate Convention. Greenpeace International, 19–28 June.

Group of Lisbon (1995), *Limits to Competition*, Cambridge, Mass./London: MIT Press.

Haas, P.M. (1993), 'Epistemic Communities and the Dynamics of International Environmental Co-operation', in V. Rittberger (ed.), *Regime Theory and International Relations*, Oxford: Clarendon Press, pp. 168–201.

Held, D. (1995), *Democracy and the Global Order. From the Modern State to Cosmopolitan Governance*, Cambridge: Polity.

Hurrell, A. and B. Kingsbury (eds) (1992), *The International Politics of the Environment. Actors, Interests, and Institutions*, Oxford: Clarendon Press.

Jacobson, H. (1984), *Networks of Interdependence. International Organizations and the Global Political System*, New York: Alfred A. Knopf.

Jänicke, M. (1993), 'Über ökologische und politieke Modernisierungen', *Zeitschrift für Umweltpolitik und Umweltrecht*, **2**, 159–75.

Kickert, W., E. Klijn and J. Koppejan (eds) (1997), *Managing Complex Networks. Strategies for the Public Sector*, London: Sage

Kohler-Koch, B. and R. Eising (eds) (1999), *The Transformation of Governance in the European Union*, London: Routledge.

Kolk, A. (1999), 'Multinationale Ondernemingen en Internationaal

Klimaatbeleid', *Milieu*, **14**(4), 181–91.

Kooiman, J. (ed.) (1993), *Modern Governance: New Government–Society Interactions*, London: Sage.

Mazey, S. and J. Richardson (1992), 'Environmental Groups and the European Community: Challenges and Opportunities', *Environmental Politics*, **1**(4), 109–28.

Miller, M.A.L. (1995), *The Third World in Global Environmental Politics*, Buckingham: Open University Press.

Mintzer, I.M. and J.A. Leonard (eds) (1994), *Negotiating Climate Change. The Inside Story of the Rio Convention*, Cambridge: Cambridge University Press.

Mol, A.P.J. (2001), *Globalization and Environmental Reform. The Ecological Modernization of the Global Economy*, Cambridge, Mass.: MIT Press.

Oberthür, S. and H.E. Ott (1999), *The Kyoto Protocol, International Climate Change Policy for the 21st Century*, Berlin/Heidelberg/New York: Springer.

OECD (1998), *OECD Information Paper: Ensuring Compliance with a Global Climate Change Agreement*, Paris: OECD.

Ohmae, K. (1995), *The End of the Nation-State*, New York: Free Press.

Ott, H.E. (1997), *Outline of EU Climate Policy in the FCCC – Explaining the EU-bubble*, Tokyo: Japan Center of International and Comparative Environmental Law.

Ott, H.E. and S. Oberthür (1999), *Breaking the Impasse: Forging an EU Leadership Initiative on Climate Change*, Berlin: Heinrich-Böll Foundation.

Princen, T. and M. Finger (eds) (1994), *Environmental NGOs in World Politics. Linking the Global and the Local*, London: Routledge.

Ritzer, G. (1993), *The McDonaldization of Society: An Investigation into the Changing Character of Contemporary Social Life*, London: Sage.

Robertson, R. (1992), *Globalization: Social Theory and Global Culture*, London: Sage.

Rolfe, C. (1998), *Kyoto Protocol to the UNFCC: A Guide to the Protocol and Analysis of its Effectiveness*, West Coast Environmental Law Association.

Telesetsky, A. (1999), 'International Law Treaties: the Kyoto Protocol', *Ecology Law Quarterly*, **1999**, 797–813.

UNEP (2000), *Report of the Working Group on Compliance and Enforcement of Environmental Conventions*, Nairobi: UNEP.

Van Kersbergen, K., R.H. Lieshout and G. Lock (1999), *Expansion and Fragmentation. Internationalization, Political Change and the Transformation of the Nation-State*, Amsterdam: Amsterdam University Press.

Van Noort, W., L. Huberts and L. Rademakers (1987), *Protest en Pressie. Een systematische analyse van collectieve actie*, Assen: Van Gorcum.

Van Tatenhove, J. (1993), *Milieubeleid onder dak*, Wageningen: Wageningen University (PhD thesis).

Van Tatenhove, J., B. Arts and P. Leroy (eds) (2000), *Political Modernisation and the Environment. The Renewal of Environmental Policy Arrangements*. Dordrecht: Kluwer Academic Publishers.

Vogel, K. (1992), 'Administrative Law, International Aspects', in *Encyclopedia of Public International Law*, Volume I.

Werksman, J. (1996), *Designing a Compliance System for the UNFCCC*, London: FIELD.

Werksman, J. (1998), *Responding to Non-Compliance under the Climate Change Regime*, OECD Information Paper, Paris: OECD.

PART II

Towards Solutions and Consequences of these Solutions

6. Policies and measures in international climate policy: price vs. quantity

Catrinus J. Jepma and Jan Bandsma

1 INTRODUCTION

An individual country can choose from a variety of international policies and measures to reduce its emissions of greenhouse gases, in the framework of the Kyoto Protocol or otherwise (see details in Chapter 1, above). Usually these instruments are conceptually divided into two categories: market-based instruments and regulatory (predominantly command-and-control) instruments. During the past decades, economists have consistently argued in favour of market-based instruments, basically because such instruments leave the decisions as well as the choices of the best environmental solutions to the private sector, which is supposedly the sector that is the best informed and the best capable of reducing its own environmental impact. The alternative system, command-and-control, which leaves the initiative for environmental regulation with the government, would suffer from the information asymmetry between government and private sector, that is from the fact that the government does not have the best and most recent information on behavioural and technological development in the private sector. In addition, command-and-control would not benefit from the flexibility of the private sector to respond to incentives in the most cost-effective manner. In short, economists tend to agree with the general statement that market-based instruments are generally superior to command-and-control in environmental policy design.

Obviously, it is not difficult to think of particular circumstances where command-and-control would still be the best way to achieve environmental targets. First of all, it is clear that a number of criteria will determine which policy instrument is the preferred one to deal with a particular environmental problem. The most obvious criteria being: whether the instrument reaches the environmental goal at minimum cost (efficiency/cost-effectiveness); whether the environmental goal is reached at all (effectiveness); whether no adverse distributional implications emerge from the policy instrument (distributive impact); and whether the implementation of the instrument raises transaction costs by too much for one reason or another (acceptability). If both a command-and-control measure and a market-based instrument are feasible to try to deal with a particular environmental issue, the market-based instrument may score better with regard to cost-effectiveness and acceptability, but the command-and-control measure would deal better with regard to effectiveness and distributive impact. Under such circumstances, in the absence of any objective system to weigh one criterion against another, it is simply impossible to assess which instrument would be the preferred one. The general inclination of economists to focus on cost-effectiveness and acceptability may, therefore, create an economists' bias in favour of market-based instruments. In actual practice it is fair to say that, despite the persistent claim of many economists that market-based instruments are the preferred ones in many circumstances in environmental policy, the overwhelming majority of environmental policy instruments are still of the command-and-control type.

The debate on the relative merits or demerits of the various policies and measures in environmental management is further complicated because so little systematic knowledge is available on the impact of various types of policies and measures on technological development. It may well be that, under certain circumstances, a well-balanced and well-chosen package of command-and-control measures is more successful in speeding up environmentally friendly new technologies than any of the market-based instruments. Another complexity occurs when the private sector is characterized by a non-competitive environment. The issue for the government to put pressure on that sector to introduce more environmentally benign production processes can become quite complex so that it is not altogether clear beforehand which instrument may be considered superior according to the various criteria mentioned.

Having said all this, we will now leave the issue of command-and-control versus market-based instruments aside, as there is already an extensive body of literature covering this fundamental issue of policy design. Another argument for leaving this issue aside in this chapter is that the same debate increasingly tends to become a discussion between believers and non-believers, whereby the weights of the various criteria are the implicit rather than the explicit arguments. Instead, in this chapter we will focus on the more topical issue of the range of market-based instruments that are now considered in the actual policy arena, and in particular on the question of which of these market-based instruments are most likely to be more widely implemented in the future climate policy arena. Conceptually, the most important distinction within the class of market-based instruments is that between policy instruments which are based on price-setting and those based on quantity-setting. In the first case, the government (or regulator) sets the price, for instance via a tax per unit of pollution, and leaves it to the individual producer to figure out what the optimal production volume would be; in the second case, the producer faces a prescribed, maximum volume of pollution, but is free to determine the optimal price at which the related product should be offered to the market. The first to make a comparative study of these two types of instruments was Weitzman (1974), later followed by Pizer (1997) and others. As we will briefly show later on, these authors basically concluded that which of the two approaches would be the preferred one under conditions of uncertainty depends on the elasticities of the marginal damage and marginal cost curves.

The debate on the pros and cons of the various market-based instruments is rather fierce nowadays in the context of the design of global climate policy. This is fairly natural, since the design of climate policy will clearly be amongst the key policy challenges, at least in the Western world, for the rest of the decade. Given what is at stake, and given the overwhelming importance of energy in virtually every sector of industrialized economies, it is no surprise that the struggle to find a good portfolio of climate policy instruments and measures that satisfies the criteria previously mentioned without undermining a level playing field in international competition or inviting free-riding is a tremendous challenge. This challenge was given a new impulse at the second session of the sixth Conference of Parties in Bonn in July, 2001 (COP6*bis*), when basically political agreement was reached (with the exception of the USA) on international climate policy design.

One could argue that the question of what type of instruments should be used actually presents itself at two levels: first, what will the overall climate policy design look like (that is, for all committed parties together) and second, how will individual parties deal with their national commitments? Obviously, the policy design issue is most relevant at the first level, if only because individual countries can decide for themselves how to deal with their commitments, and, therefore, not necessarily in a uniform manner. This makes it all the more striking that there never seems to have been an intensive debate on the optimal design of a global climate policy, either among scientists or among policymakers or both. All of a sudden, during COP3 in Kyoto, the decision was made to set up a global system of committed and non-committed parties, whereby committed parties basically agreed to accept a cap-and-trade system. Any alternative system, for instance, one based on internationally harmonized taxation, or well-coordinated policies and measures through some kind of common action programme, does not seem to have been evaluated, assessed or even seriously considered in any systematic manner.

Actually, the Kyoto Protocol (KP) design, based on country commitments together with the trading option between Annex I countries (Article 17 KP) and with the project-based mechanisms (JI and CDM, Articles 6 and 12 KP) was welcomed as such a major breakthrough that considering any alternative systems simply appeared pointless. At first glance, the design of the KP indeed looks very attractive from many points of view. However, subsequent history has shown that the elements of the cap-and-trade system on which the KP is based have actually produced considerable difficulties in the subsequent negotiations. It appears that the key insights that had been derived from theory on the applicability of price-based versus quantity-based were almost completely ignored in the overall design of the Kyoto Protocol.

Therefore, in this chapter we will first focus on some of the key lessons that can be learned from theory, not because they are particularly new but because it may be particularly relevant to reconsider that theory once again in the light of the Kyoto Protocol design (Section 2). We will then try to make a broad assessment of the Protocol design against the background of what theory has to say about market-based instruments (Section 3). Finally, we will turn to the national level, and focus on the debate that has started quite recently about the question of the circumstances under which a tax or some trading system would be the preferred option (Section 4). Section 5 will present some conclusions.

2   PRICE-BASED VS. QUANTITY-BASED INSTRUMENTS: THEORETICAL ANALYSES

Weitzman's work entitled *Prices vs. Quantities* (1974) was the first comparative analysis of price-based and quantity-based instruments. He presented the case of a government that wants to regulate a market and has two instruments at its disposal: a price (tax) and a quantity (permit) instrument. The question then posed is which instrument is preferable from a welfare point of view. When there is no uncertainty about the cost, there is no preference for either instrument: at each price, there is a known associated quantity where the price equals marginal cost. This is shown in Figure 6.1 at the point where the social marginal costs curve (smc) intersects with the marginal benefit curve (mb). At this point the socially optimal pollution control level is chosen. Since there is perfect information, we can see that it does not matter whether this optimum is achieved by a tax (t_1) or by setting a cap on volume (X_1). In the case of a tax, the social optimum is reached by raising the marginal cost (mc) to the level of t_1 so that social marginal costs equal social marginal benefits. In the case of a cap system, the amount of pollution is simply set at X_1, the socially optimum level.

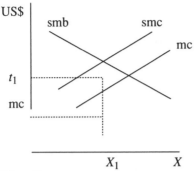

Source: Hanley et al. (1996), p. 59.

Figure 6.1 Socially optimal level of pollution control

However, when the cost is uncertain, fixing the price at a certain level with a tax will lead to an uncertain quantity level. And *vice versa,* fixing the quantity at a certain level will lead to uncertainty about the price of abatement units, and therefore about the total cost of climate policy. Basically, the result of Weitzman's analysis is that when the marginal benefit (or damage) function is relatively flat, a price instrument would be preferable. Certainty about the price would then be preferred over certainty about the quantity, since the difference in levels of damage corresponding with the difference in levels of quantity is relatively small. Conversely, when the marginal cost curve is relatively flat, a quantity-based instrument is preferred. So the welfare loss is minimized by choosing the type of policy instrument that results in outcomes that are closer to the outcomes in the case of total certainty than the next best instrument.

Roberts and Spence (1976) were the first to suggest the use of a hybrid policy mechanism. In such a system the regulated industry would have the choice of either purchasing a permit in the marketplace or from the government at a certain fixed price (the trigger price). If the market price lies below the trigger price, permits will be bought in the marketplace. When the price rises above the trigger price, permits will be bought from the government and can then be regarded as a kind of tax. The question is where to set the trigger price. When the trigger price is set at a sufficiently high level, it will function as a permit system, and conversely when the trigger price is set low enough, it will function as a tax system. The advantage of a hybrid system is that, from a welfare point of view, it will be at least as good as either a tax system or a permit system, so that the most efficient instrument will automatically function. Such a hybrid system, however, fails to guarantee particular emission levels, but does limit the economic cost of the programme for its users (IPCC, 2001, p. 417).

In the context of the current climate change debate and the Kyoto Protocol, McKibben and Wilcoxen (1997) argue that a hybrid policy would lower costs, improve monitoring and enforcement, and avoid disruptive international capital flows. The greenhouse gas emission quota as stated in Annex B of the Kyoto Protocol would then not be strict emission limits but emission targets that would be met as long as the cost of emission reduction does not exceed a certain level. Pizer (1997) analyses the tax instrument, permit system and hybrid system in relation to climate change policy with uncertainty about the cost of emissions reduction. The result of this analysis is that from a welfare point of view taxes are much more (by a factor of five) efficient than permits, which is the result of a relatively flat marginal benefits (damage) curve. It also shows that the hybrid system is slightly better than the tax system.

To summarize, in determining whether price-based or quantity-based instruments would be a more favourable alternative, the degree of uncertainty and the correlation between the marginal damage and marginal cost curves play decisive roles. Where there is no clear, short-term threat of severe climate-related damage, the use of price-based policies, like taxes, seem to be preferable since they limit cost uncertainty.

3   THE KYOTO PROTOCOL

The Kyoto Protocol design is clearly based on the quantity-based, cap-and-trade approach, the cap being the commitment specified per party (at least for the first commitment period) and the trading element being the three so-called Kyoto mechanisms (JI, CDM and IET). At first glance, the approach looks rather attractive: the cap ensures that the predetermined environmental target will be reached, whereas trading makes it possible to reach that goal as cost-effectively as possible. The supposedly 'fair' distribution of the 'cap' over the parties, finally, would guarantee political acceptability along with an acceptable distributive impact.

In actual practice this general design, attractive as it initially looks as an abstraction, has proved to be much more complex and subject to debate than some would have thought beforehand. That is despite the fact that the Kyoto Protocol, together with its design, was accepted surprisingly swiftly at the Third Conference of the Parties (COP3) in Kyoto in 1997.

What makes the concept so questionable? First, the predetermined environmental target itself has no clear basis in science at all. Those who are familiar with climate modelling work and climate scenarios are aware that the various scenarios show a widely varying pattern as far as the projected emissions in the period until 2100 is concerned. But even (a) if some feasible range were to be singled out – on whatever criterion – as being the most likely estimation of an emissions range, along with (b) some stabilization target of greenhouse gas concentrations in the atmosphere 'at a level that would prevent dangerous anthropogenic interference with the climate system' (UNFCCC, Article 2) of, say, 550 ppmv, then still it is rather arbitrary to translate that data into a clear target for the 2008–2012 first commitment period, which is clearly no more than just a first short period in a supposedly long policy action period. As a result, it is very unlikely that all Annex I Parties would be satisfied with the compromise finally agreed (–5.2 per cent *vis-à-vis* 1990 emission levels for the total Annex I area, and a 10 to 20 per cent reduction from their projected emissions during 2008–2012). Some parties or specific groups, such as 'green' NGOs, regard this as far too lenient, while others are of the opinion that it puts an unnecessary strain on the political and economic system, the US government being a clear example of the latter.

What makes the Kyoto Protocol target even more questionable is that – precisely because it in fact sets a cap – it is crucial that the actual emissions per party are measured precisely against that cap. That is because if the difference is positive (emissions > cap), this will imply a proportionate emission penalty (1.3 deduction rate for the first commitment period) for that party. In actual practice, however, the national measurement systems for the various greenhouse gas distinguished in the Protocol have thus far only been poorly developed, using many imprecise assumptions and 'guesstimates'. This is obvious since the greenhouse gas emissions reporting obligations at the sector and firm level from which the national emissions data have to be compiled have thus far not been properly developed either. As a result, the margins of error of national greenhouse gas emissions for most Western countries are most likely well in excess of 5 per cent or even 10 per cent of the annual data, especially given the imprecision of the data for non-CO_2 emissions. The result is somewhat confusing because for most parties the margins of error of their national emissions (average expected value, aev) may well actually prove to be many times greater than the degree to which the same parties' emissions deviate from the cap. So, to illustrate with the help of a simple example, a cap of 100 with actual emissions (aev) of 98, would suggest a surplus of two. If, however, the margin of error is 10, the actual situation may range from a surplus of 12 to a deficit of 8![1]

What makes the system precarious as well is that for rather arbitrary reasons it was decided to measure the emissions on the production side rather than on the side of the final consumer (who would eventually need to pay).

Under those circumstances it pays to relocate energy-intensive production facilities to other countries, because the Kyoto Protocol mechanism rewards such behaviour with a premium. The international relocation bias as such would not be so problematic if the industries moved to other committed areas: the new areas would then have to do something extra in return for receiving these foreign industries. The problem, however, is that this zero-sum game does not apply if firms move to the non-committed non-Annex area. This will result in carbon leakage by creating a loophole in the system that, if used effectively, waters down the predetermined environmental target.

Even more peculiar and controversial than the target itself is the way in which caps are distributed among the parties. First, the dividing line between committed Annex and uncommitted non-Annex Parties is a rather arbitrary one, which should certainly not be static. But even given this dividing line, the question of how caps are split up and assigned to the various Annex Parties is a difficult one, simply because most countries realize that they can best serve their own interests by arguing that they need a favourable exception for themselves for one reason or another. The acceptance of the actual distribution, as well as the EU-bubble-based distribution, during the Kyoto negotiations came as a pleasant surprise. One should not forget, however, that at that time parties still had the option – if the impact of their cap impact proved too severe – not to ratify or, if ratification was considered, not to accept a tough compliance regime, and/or to see to it that sufficient loopholes were created to relax the cap's impact. Now already (October 2001), the USA has withdrawn from the Kyoto Protocol; parties such as Japan, Australia and Canada have indicated that their ratification is conditional on the USA joining the Protocol in the future; and the story goes that individual EU countries want to re-open discussions on the bubble soon, at least before the start of the first commitment period.

The most complex part of the system has thus far turned out to be the trading element. Although some key issues such as a supplementarity rule and an acceptable compliance regime seem to have been solved in the meantime, a significant number of issues regarding trading still remains to be solved. What design characteristics of the JI and CDM mechanisms would make them attractive to all parties and entities involved? What would the role of government and private sector need to be? What about liability and compliance, or the consequences of non-compliance? It is no use spelling out all the complexities here, but it is in any case clear that the part of the cap-and-trade system most likely to be attractive in theory, that is the trading part (as opposed to the commitment allocation part), has in practice turned out to be a negotiators' nightmare. The odd thing is that the only reason why a cap-and-trade system, such as emissions trading, is called market-based at all is its trading aspect; all the other elements – the cap and its allocation (unless through auction) – have in fact nothing to do with the market mechanism at all. Yet it is the trading element in the Kyoto Protocol that up to now seems to have created even more trouble than the cap-and-allocation element.

To return to the theoretical analysis, since the Kyoto Protocol is provided with a rather weak compliance and inventory system the notion of it being a cap-and-trade system is a bit less true. The penalty for not being in compliance after the first commitment period is simply that such a party has to carry over this part of its assigned amount deficit (that is surplus in emissions) to the second commitment period with an extra penalty of a 1.3 deduction rate. This is equivalent to what is known in the literature as 'borrowing' from a future commitment period. However, the assigned amounts per party for the second commitment period could be negotiated rather late, and theoretically even after the end of the first commitment period. This would mean that the party that was not in compliance during the first commitment period will take into account this extra effort later on when negotiating its assigned amount for the second commitment period. Thus, the 'penalty' from the first commitment period may well evaporate by adjusting at least part of the terms of the commitment to be decided in the negotiating process preceding the second commitment period.

4 OPTIMAL CLIMATE POLICY

As mantioned before, the UNFCCC and Kyoto Protocol are essentially quantity-based in their design, the quantified emissions limits being the commitments allocated among the parties listed in Annex B. This implies that, with greenhouse gas emission limits for the six relevant gases mentioned in the Kyoto Protocol being certain, there will be uncertainty about the total cost of abatement not only globally, but also for each country.

A significant number of economists seem to agree that market-based approaches are preferable over command-and-control, simply because they generally offer the best prospects of achieving cost-effectiveness. Given that climate policies will meet political resistance anyway, any reduction in the costs of policy packages for society should therefore be welcomed.

The question of which policy instruments are the preferred ones, and in particular what market-based instruments should be set up to try to deal with the Kyoto commitments, arises not only at the global but also at the national level. It is clear that there is no *a priori* reason why what is best at the global level would also apply at the level of all individual parties. In any case, parties have at least a considerable degree of freedom to implement different measures to comply with their quantitative emission targets. First of all, they have the range of domestic non-market-based policies and measures (see Box 6.1). Second, they have the domestic market-based policies and measures, which could be quantitative-based (for example a domestic emissions trading scheme) or price-based (for example taxes). Finally, they can make use of the international policies and measures (see Box 6.2), including the Kyoto mechanisms (JI, Article 6; CDM, Article 12; and IET, Article 17). A country will most likely use a mix of various policy instruments rather than just one or two instruments, as most theoretical models would assume. It may well be that the various policy criteria mentioned in the Introduction to this chapter can only be satisfied simultaneously with a variety of policies and measures.

BOX 6.1   DEFINITIONS OF SELECTED NATIONAL GREENHOUSE GAS ABATEMENT POLICIES AND MEASURES

- An **emissions tax** is a levy imposed by a government on each unit of CO_2-equivalent emissions by a source subject to the tax. Since virtually all of the carbon in fossil fuels is ultimately emitted as CO_2, a levy on the carbon content of fossil fuels – a carbon tax – is equivalent to an emissions tax for emissions due to fossil fuel combustion. An energy tax – a levy on the energy content of fuels – reduces demand for energy and so reduces CO_2 emissions due to fossil fuel use.
- A **tradable permit (cap-and-trade) system** establishes a limit on aggregate emissions by specified sources, requires each source to hold permits equal to its actual emissions, and allows permits to be traded among sources. This is different from a credit system, in which credits are created when a source reduces its emissions below a baseline equal to an estimate of what they would have been in the absence of the emissions reduction action. A source subject to an emissions limitation commitment can use credits to meet its obligation.
- A **subsidy** is a direct payment from the government to an entity, or a tax reduction to that entity, for implementing a practice the government wishes to encourage. Greenhouse gas emissions can be reduced by lowering existing subsidies that have the effect of raising emissions, such as subsidies to fossil fuel use, or by

providing subsidies for practices that reduce emissions or enhance sinks (for example for insulation of buildings or planting trees).

- A **deposit–refund system** combines a deposit or fee (tax) on a commodity with a refund or rebate (subsidy) for implementation of a specified action.

- A **voluntary agreement (VA)** is an agreement between the government and one or more sources to limit greenhouse gas emissions or to implement measures that will have that effect. The agreement may allow participants to trade emission reduction credits among themselves (or with participants in other agreements) to reduce the cost of compliance.

- A **non-tradable permit system** establishes a limit on the greenhouse gas emissions of each regulated source. Each source must keep its actual emissions below its own limit; trading among sources is not permitted.

- A **technology or performance standard** establishes minimum requirements for affected products or processes that reduce greenhouse gas emissions associated with the manufacture or use of the products or process.

- A **product ban** prohibits the use of a specified product in a particular application, such as hydroflourocarbons (HFCs) in refrigeration systems, that give rise to greenhouse gas emissions.

- **Direct government spending and investment** involves government expenditures on R&D measures that will lower greenhouse gas emissions or enhance sinks.

The first four policy instruments in the list above are often called market-based instruments, although some voluntary agreements can also fall into this category.

Source: IPCC (2001), p. 404.

BOX 6.2 DEFINITIONS OF SELECTED INTER-NATIONAL GREENHOUSE GAS ABATEMENT POLICIES AND MEASURES

- A **tradable quota system** establishes a national emissions limit for each participating country and requires each country to hold quota equal to its actual emissions. Governments, and possibly legal entities, of participating countries are allowed to trade quotas. Emissions trading under Article 17 of the Kyoto Protocol is a tradable quota system based on the assigned amounts calculated from the emission reduction and limitation commitments listed in Annex B of the Protocol.
- **Joint implementation** (JI) under Article 6 of the Kyoto Protocol allows the government of, or entities from, a country with a greenhouse gas emissions limit to contribute to the implementation of a project to reduce emissions, or enhance sinks, in another country with a national commitment and to receive credits equal to part, or all, of the emission reduction achieved. The credits can be used by the investor country or other Annex I Party to help meet its national emissions limitation commitment. Article 6 of the Kyoto Protocol establishes JI among parties with emission reduction and limitation commitments listed in Annex B of the Protocol.
- The **clean development mechanism** (CDM; JI under Article 12 of the Kyoto Protocol) allows the government of, or entities from, a country with a greenhouse gas emissions limit to contribute to the implementation of a project to reduce emissions, or possibly enhance sinks, in a country without a national commitment and to receive credits equal to part, or all, of the emission reduction achieved. Article 12 of the Kyoto Protocol establishes the CDM to contribute to sustainable development of the host country and to help Annex I Parties meet their emission reduction and limitation commitments.
- A **harmonized emissions/carbon/energy tax** commits participating countries to impose a tax at a common rate on the same sources. Each country can retain the tax revenue it collects.
- An **international emissions/carbon/energy tax** is a tax imposed on specified sources in participating countries by an international agency. The revenue is distributed or used as specified by participating countries or the international agency.

- **Non-tradable quotas** impose a limit on the national greenhouse gas emissions of each participating country to be attained exclusively through domestic actions.
- **International product/technology standards** establish minimum requirements for affected products/technologies in countries where they are adopted. The standards reduce greenhouse gas emissions associated with the manufacture or use of the products/application of the technology.
- An **international voluntary agreement** is an agreement between two or more governments and one or more sources to limit greenhouse gas emissions or to implement measures that will have that effect.
- Direct **international transfers of financial resources and technology** involve transfers of financial resources from a national government to the government or legal entity in another country directly or via an international agency with the objective of stimulating greenhouse gas emission reduction or sink enhancement actions in the recipient country.

The first five policy instruments in the list above are often called market-based instruments, although some voluntary agreements can also fall into this category.

Source: IPCC (2001), p. 405.

So what would the Weitzman et al. theory discussed in 6.2 suggest as the optimal policy mix for a country listed in Annex I given the design of the Kyoto Protocol after the Bonn agreement? The answer depends greatly on the actual compliance regime. Since it can be argued on the one hand that the Kyoto Protocol compliance/inventory regime probably will be rather soft, at least during the first commitment period, and since on the other the caps will not be easy to satisfy for most parties, Weitzman's model would suggest that it does not seem very logical for a country to choose a quantitative-based domestic policy instrument, except for those countries that can easily attain their cap. One would instead expect countries to opt for price-based systems, such as taxation, combined with various other instruments, including the Kyoto mechanisms.

The debate between the proponents of *quantity-based* (emissions trading) vs. *price-based* (taxation) instruments is still quite open. On the whole, the

Anglo-Saxon world slightly favours the former, and the rest of the industrialized world slightly favours the latter instrument. Similarly, there has been some controversy in the academic literature on the issue.

In the first half of the 1990s most emphasis seems to have been on Pigovian taxes and the double-dividend issue; at the same time many analyses have been carried out within the EU on the implementation of European carbon taxes. Double-dividend occurs when a climate policy both improves the environment and offsets at least part of the welfare losses of climate policies by reducing the costs of the tax system. Factors that determine the magnitude of the double-dividend are the direct cost to the regulated sector, the tax interaction effect and the revenue-recycling effect (IPCC, 2001, p. 513). In general, modelling results show that the sum of the positive revenue-recycling effect and the negative tax-interaction effect can be both positive and negative. In economies with an especially distorting tax system (as in several European analyses), the sum may be clearly positive (resulting in a so-called strong double-dividend effect). In various models of the US economy, the sum turned out to be negative (a so-called weak double-dividend effect). Analyses of the US economy found that revenue recycling reduces the cost of regulation by between about 30 and 50 per cent for a certain range of targets, while European analyses report cost savings that are even higher than 100 per cent (IPCC, 2001, p. 521).

However, in the second half of the 1990s the emission trading schemes became fashionable, and the issue of cost-effectiveness and trading modalities received a lot of attention. Why would countries currently start designing a domestic emissions trading regime, as in the European Union, or even implementing one, as in the case of the UK? First of all, there is clearly some pressure, especially in Anglo-Saxon neo-classical academic circles, to start implementing the cap-and-trade instruments. The American SO_2 trading system experience also helps to illustrate how the instrument can work in practice.[2] This may not come as a surprise since emission-trading schemes combine the two strong characteristics of achieving the predetermined environmental target at low costs. So the criteria of efficiency and environmental effectiveness are clearly met. Since (neo-classical) economists typically tend to put a heavy emphasis on those two criteria, it stands to reason that emission trading receives so much attention.

At the same time, emissions trading may score rather badly as far as the criteria of distributional impact, implementability and transaction costs are concerned, criteria which may not always receive the necessary attention in academic circles. Particularly the issues of permit allocation, the required, rather detailed, measurement systems, the requirement of a fair and transparent trading system and of a reliable liability/compliance regime are responsible for the general poor score on the latter criteria. One can argue that transaction costs and pitfalls that accompany a cap-and-trade system are easily underestimated. Transaction costs involving the continuous monitoring, registration and verification of the emissions per firm; the establishment and operation of a national registry that keeps track of all the transactions of greenhouse gas allowances; the 'fair' allocation of the permits to the individual firms, while keeping in mind the effect it will have on domestic and foreign competition in that sector; how to enforce a compliance regime without seriously damaging the sector; the problem of speculation in the market; these are some of the issues that may raise serious obstacles and costs once cap-and-trade systems are implemented in practice.

After the wave of publications in favour of the cap-and-trade instrument in the second half of the 1990s, the emphasis seems to be shifting back to the taxation instrument, particularly in policy-oriented circles. According to the IMF (2000), in the policy arena tradable permits are still somewhat controversial, even in the United States, where they are probably most widely propagated and most extensively used. Based on an extensive inventory, the IMF argues that the vast majority of countries indicated that they favoured taxes over tradable permits for pollution control (IMF, 2000), presumably because taxes can also be implemented through an existing administrative apparatus. The leading countries in the use of eco-taxes found taxes targeting emissions of carbon dioxide to be the most efficient way of reaching climate policy targets. A recent OECD (2001) report also concludes that there is growing evidence that such environmental taxes actually work very well in practice.

On the whole, the appetite for carbon taxation seems to be growing. In practice, both energy and carbon taxes have already been adopted as responses to commitments under the UNFCCC. The European Commission has issued several tax proposals designed to reduce emissions of carbon dioxide from fossil fuel use. Finland, the Netherlands, Denmark, Sweden and Norway, to name just a few, all have energy taxes based in part on carbon content (Speck, 1999). Other countries that have recently introduced carbon or energy taxes to help achieve their climate change commitments include Slovenia, the UK, Italy, Germany and Switzerland. France is also thinking about increasing energy taxes on industry for that same purpose. None of these countries have been able to introduce a uniform carbon tax for all fuels in all sectors, mainly because of the unilateral nature of their policy and concerns expressed by industry exposed to international competition. In most cases where an energy or carbon tax is implemented, the tax is implemented in combination with various forms of exemptions (for example, rebates, voluntary agreements) (IPCC, 2001, pp. 414–15).

Other new arguments in favour of taxation based on theory have been presented recently. An MIT study (Fullerton and Metcalf, 2001), for instance, argued that the introduction of a cap-and-trade programme in a market characterized by imperfect competition is more likely to reduce rather than increase welfare, even if the environmental benefits are taken into account.

Given the various perspectives highlighted in the literature, it is not easy to predict how the various future regimes to fulfil the Parties' commitments will appear. It may well be that Weitzman's lesson will be learned, and that taxation rather than emission trading will develop into the core domestic instrument in dealing with the Protocol. The debates between governments and industry on the principles of permit allocation, exemption regimes for reasons of international competitiveness and costs, and trading participation have been rather complex thus far, and have generally tended to substantially water down the overall environmental targets of emission trading schemes. The latter point also applies to proposals for hybrid trading schemes (IPCC, 2001, p. 417). The latest design of the EU-wide emissions trading scheme, which is scheduled to start in 2005, has incorporated the possibility for governments to put extra permits into the market if the price per tonne of CO_2 equivalent becomes too high (Reuters, 2001). All this, along with the massive costs of emissions measurement at the industry level and the impossibility to get the required precision in measurement, as well as the fear that trading may be manipulated by speculators who succeed in raising permit prices to their advantage, may eventually turn out to be stumbling blocks for the large-scale introduction of such schemes.

The backbone of domestic policies and measures for climate policies, therefore, may well eventually turn out to be carbon taxation. However, taxation alone would most likely not be sufficient to accomplish the entire emission reduction obligation either; in any case parties may not be certain that their commitments are fulfilled through domestic taxation. Since countries would probably not want to face the risk of not meeting their commitments at a rather late stage, they may want to create a 'safety zone' for its emissions at an earlier stage. Such a 'safety zone' could actually be achieved via a host of other instruments, including the Kyoto mechanisms.

5   CONCLUSIONS

The classical theory of price-based and quantity-based policy design suggests, given the weak compliance and inventory regime of the Kyoto Protocol, that a price-based policy (taxes) would seem to be the preferred one. In the actual practice of implementing domestic climate policy, taxes may also eventually turn out to be the preferred instrument, and thus the backbone of domestic policies and measures to meet the Kyoto Protocol commitments. For one thing, it is probably the least market-disruptive instrument and it gives those being charged a considerable amount of certainty about the costs to them of climate policy. Second, taxes perform relatively well in terms of transaction costs and ease of implementation. Moreover, they do not suffer from the problem of permit allocation, which is typically associated with the quantity-based emissions trading regime, or the risk of abuse of the permit market through speculation. Finally, taxes have the advantage that governments are familiar with that instrument. The challenge may well be to combine them with additional policies and measures, including the Kyoto mechanisms.

NOTES

1. See *Energy & Environment*, Special Issue (2002, forthcoming).
2. Tradable permits have been used to implement a cap on SO_2 emissions by electricity generators in the USA and on NO_x and SO_x emissions by large sources in the greater Los Angeles area (the RECLAIM program) (see for instance Schmalensee et al., 1998.

REFERENCES

Energy & Environment (2002), *Special Issue – Climate Change: Mitigation, Measurement and Measures*, UK: Multi Science Publishing Co. Ltd, forthcoming.

Fullerton, D. and G.E. Metcalf (2001), *Cap and Trade Policies in the Presence of Monopoly and Distortionary Taxation*, Cambridge, USA: Massachusetts Institute of Technology.

Hanley, N., J.F. Shogren and B. White (1996), *Environmental Economics in Theory and Practice*, London, UK: MacMillan Press Ltd, London.

IMF (2000), *Controlling Pollution-Using Taxes and Tradable Permits*, Washington, DC: International Monetary Fund.

IPCC (2001), *Climate Change 2001: Mitigation – Contribution of Working Group III to the Third Assessment Report of the Intergovernmental Panel on Climate Change*, UK: Cambridge University Press.

McKibbin, W.J. and P.J. Wilcoxen (1997), 'A Better Way to Slow Global Climate Change', *Brookings Policy Brief*, **17**, Washington DC: The Brookings Institution.

OECD (2001), *Environmentally Related Taxes in OECD Countries: Issues and Strategies*, Paris: Organisation for Economic Co-operation and Development.

Pizer, W.A. (1997), 'Prices vs. Quantities Revisited: The Case of Climate Change', *Resources for the Future*, Discussion Paper, **98-02**, Washington, DC: RFF.

Reuters (2001), *EU Eases CO$_2$ Trading Rules after Industry Pressure*, UK: Reuters News Service, Internet: www.planetark.org/dailynewsstory.cfm/newsid/12613.story.htm.

Roberts, M.J. and M. Spence (1976), 'Effluent Charges and Licenses under Uncertainty', *Journal of Public Economics*, **5**, 193–208.

Schmalensee, R., P.L. Joskow, A.D. Ellerman, J.P. Montero and E.M. Bailey (1998), 'An Interim Evaluation of Sulfur Dioxide Emissions Trading', *Journal of Economic Perspectives*, **12**(3), 53–68.

Speck, S. (1999), *A Database of Environmental Taxes and Charges*, European Union, DG Environment.

UNFCCC (1992), *Framework Convention on Climate Change,* Bonn: UNFCCC Secretariat.

Weitzman, M. (1974), 'Prices vs. Quantities', *Review of Economic Studies*, **41**, 477–91.

7. The Kyoto mechanisms: economic potential, environmental problems and political barriers

Edwin Woerdman, Wytze van der Gaast, Ton Manders and Andries Nentjes

1   INTRODUCTION

The Kyoto Protocol allows industrialized countries – the so-called 'Annex B parties' – to fulfil their commitments to greenhouse gas abatement partly through reducing emissions in other countries where such measures are cheaper. This kind of flexibility, which enables industrialized countries to comply with their commitments at lower costs, is defined in the Protocol under the following so-called 'Kyoto mechanisms':

- joint implementation (JI) under Article 6;
- the clean development mechanism (CDM) under Article 12; and/or
- international emissions trading (IET) under Article 17.

An industrialized country can purchase either assigned amount units (AAUs) on the basis of IET or emission reduction units (ERUs) on the basis of JI projects from another Annex B country, for instance in Central or Eastern Europe, or both. It can also acquire certified emission reductions (CERs) from developing countries based on CDM projects. There are several differences between these flexible instruments.

IET uses a top-down approach by calculating reductions in emissions on the basis of national commitments. The legal text of Article 17 indicates that Annex B governments could trade parts of their assigned amounts. A sovereign government could decide to split up its assigned amounts by allocating permits to individual companies (or sectors), enabling them to trade emissions domestically. According to the Marrakesh Accords of 2001, a Party may authorize legal entities to transfer and/or acquire emissions under Article 17.

JI and the CDM differ from IET, because the former are project-based mechanisms with an investor receiving credits for the emission reductions achieved in the host country. In principle, the emission reductions in such projects are measured bottom-up from a baseline that estimates what the emissions at the project location would have been if the project had not taken place.

Although both JI and the CDM are project-based, they differ in that a JI host country has an emission target whereas a CDM host country does not. This means that there is a stronger incentive for a CDM host country to over-estimate emission reductions by inflating project baselines (in order to claim more credits). In contrast, a JI host has an assigned amount and runs the risk of being in non-compliance if it transfers too many credits. Furthermore, credits that accrue from CDM projects between 2000 and 2008 can be banked and used for the commitment period 2008 to 2012, which is not the case for JI projects. In principle, both JI and the CDM allow carbon sequestration through sink enhancement, although the use of such projects under the CDM is restricted to afforestation and reforestation only. Forest management projects that aim at protecting existing forests instead of actually (re)planting trees can be applied to a limited extent under JI, but they are not eligible as CDM projects. Moreover, the institutional and project requirements under the CDM, in terms of supporting sustainable development in the host countries (and the requirement of a supervising executive board), are stricter than under JI.

Theoretical economic models predict large cost savings from emissions trading. A market for trading carbon emissions can work well provided that it is designed adequately. In particular, it requires the participation of private entities, clear trading and enforcement rules, as well as information and trade facilities (such as a clearinghouse) to avoid market power, to strengthen compliance and to keep transaction costs low, for instance (for example Tietenberg, 1999; Michaelowa and Dutschke, 1998).

However, there are several political, environmental and institutional barriers that make it difficult to implement an efficient and environmentally effective international market for carbon trading. The basic issues of such a 'green' market are as follows: Why trade emissions? Who should trade emissions? How environmentally sound are the tradable items? How much trading should be allowed? Therefore, the objectives of this chapter are (a) to assess the theoretical cost-saving potential of the Kyoto mechanisms and (b) to analyse the barriers to implementation that correspond to these basic issues, namely:

- private-sector participation;

- baseline determination;
- hot air;
- restrictions on trade.

We use an economic approach to analyse the barriers to implementing the Kyoto mechanisms (see Chapter 6 above for a general analysis of policies and measures). The identification of these barriers and of policy options to overcome them is regarded as a priority research area in the Third Assessment Report of the IPCC (Banuri et al., 2001, p. 71).

This chapter is organized as follows: Section 2 quantifies the economic gains of using the Kyoto mechanisms, not only in terms of reducing costs, but also in terms of reducing carbon leakage. Section 3 discusses the possibilities of private-sector participation under JI, the CDM and IET. Section 4 analyses the issues of environmental integrity in baseline determination (under JI and the CDM) and hot air (under IET). Section 5 analyses the possible reasons for and consequences of a quantitative restriction on the use of the Kyoto mechanisms. Finally, in Section 6, we present our conclusions.

2 ECONOMIC GAINS OF USING THE KYOTO MECHANISMS

2.1 Cost Savings from Using the Kyoto Mechanisms

Using the Kyoto mechanisms can lead to substantial reductions in cost. The basic argument in favour of these flexibility instruments is well known and widely accepted: they allow greenhouse gas abatement to be carried out where marginal costs are lowest (for example Tietenberg, 1999; Zhang and Nentjes, 1999). To illustrate the economic gains of the Kyoto mechanisms, we consider three policy cases: no trade (NT), Annex B trading (AT) and global trading (GT). The calculations are made with the WorldScan model, which is a multi-region, multi-sector applied general-equilibrium model (CPB, 1999).

Figure 7.1 shows carbon prices (marginal abatement costs) in all three cases. For the NT case, the effects for the USA, the EU and Japan are identified separately. To capture the effect of different baselines, only ranges are given. In the NT case, no use is made of the Kyoto mechanisms and the Kyoto targets are reached within Annex B by each model region separately. The NT case shows large differences in marginal abatement costs across the regions. The WorldScan model projects marginal abatement costs for the USA, which are clearly lower than in the EU or Japan. This crucial

assumption is also in line with most (but not all) other economic models.[1]
Trading emissions by making full use of IET and JI within Annex B in the
AT case in WorldScan equalizes marginal abatement costs over the model
regions. Benefiting from low-cost options within Annex B reduces the carbon
price considerably. The withdrawal of the USA from the Protocol in March
2001 has lowered demand and decreases the carbon market price compared to
the situation that they were still in. Simulations with WorldScan show that
costs of compliance are more than halved as a result of the opt-out by the
USA. Carbon prices and associated costs fall dramatically when the GT case
is considered, by some 80 per cent compared to the AT case. This low carbon
price might indicate the potential cost savings of the CDM.

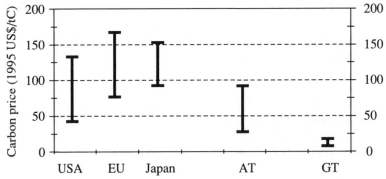

Notes:

NT = no trade.

AT = Annex B trading.

GT = global trading.

Figure 7.1   Carbon prices in three scenarios: NT, AT and GT

In theory, making full use of all Kyoto mechanisms leads to large cost
savings. However, the percentage given above is only a rough estimate, since
actual practice could be different for the following reasons. First, the
WorldScan model uses strongly aggregated cost curves for emission
reduction in predefined geographical regions that consist of several countries.
In the future, these cost curves may turn out to be correct, but they could also
be too high or too low. Second, the cost savings could be less because of
transaction costs, such as the costs of finding a trading partner. Nevertheless,
the cost savings from using the Kyoto mechanisms are still likely to be
substantial, making compliance cheaper and easier.

Greenhouse gas abatement may imply economic (welfare) costs. These costs can be measured by macro-economic consumption losses, as shown in Figure 7.2. As the scope for emissions trading increases, costs decline. In general, costs are modest, with consumption losses relative to the baseline in the order of tenths of a per cent. Not only is the share of the energy sector in Annex B economies rather small, it should also be kept in mind that the economic cost corresponds to the efficiency loss in the economy as a whole, which is clearly much lower than the total revenue of a carbon tax would be. Nevertheless, some energy-intensive sectors, such as the chemical industry, might be substantially affected.

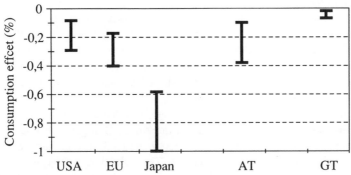

Notes:

NT = no trade.

AT = Annex B trading.

GT = global trading.

Figure 7.2 Consumption effects in three scenarios: NT, AT and GT

2.2 Carbon Leakage and the Kyoto Mechanisms

Carbon leakage refers to the situation where policies aimed at reducing greenhouse gas emissions lead to an increase in emissions in non-abating countries or in domestic regions outside a project area. In this chapter, we focus on leakage that causes an indirect emission increase, which may occur for at least two reasons: first, energy-intensive industries relocate to non-Annex B countries where energy is cheaper (the 'trade channel'). Second, lower energy prices result in more energy-intensive production processes and higher energy consumption (the 'price channel'). According to simulations with WorldScan, the leakage rate, defined as the increase in emissions in non-Annex B countries as a percentage of the reduction in Annex B countries, can be as high as 20 per cent in a situation without emissions trading.

Theoretically, carbon leakage is likely to decline with emissions trading as more flexible options reduce distortions and lower carbon prices. Compared to the NT case, the competitiveness of energy-intensive production in Annex B countries will deteriorate less, and leakage through the 'trade channel' will be lower. However, in contrast with the other Kyoto mechanisms, the CDM, particularly, could fuel carbon leakage because of the absence of a quantitative target for developing countries (which would be an implicit safeguard against leakage). The following sources of carbon leakage can be identified for the CDM.

In theory, a CDM project could increase emissions in the host country outside the project. Due to the subsidy, unit production costs of energy-intensive production could decrease, resulting in the host country's output of energy-intensive industries expanding at the cost of other countries' market share. In that case, lower energy intensity in the host country would go hand in hand with higher total energy use.

Another potential source of leakage is related to possible regional differences in energy prices. If part of the Annex B target is reached outside the Annex B area, non-Annex B energy demand will be reduced, while Annex B energy demand will increase. The reduction in energy demand in non-Annex B countries would have an effect on regional energy markets. If energy markets are not entirely globalised (due to transportation costs, for example for coal) the lower domestic energy demand would lead to lower domestic energy prices. This effect – also called 'snap-back' – would lead to leakage (more energy use) within the non-Annex B region. Investments outside the CDM projects would then become more energy intensive.

In order to limit the effects of the CDM on carbon leakage, projects could be focused on the replacement of existing capital by cleaner production technologies with the same production capacity. Such a replacement would not change total capacity in the host country, nor would it change the costs of new capacity. Although focusing on existing capital reduces carbon leakage, it cannot be completely excluded. For example, the replacement could increase future capacity if the life expectancy of the new capacity were longer than that of the old one. Leakage could also occur if investments led to technology spill-overs to other investments in the host country at no cost.

Figure 7.3 shows a WorldScan quantification of two policy cases, namely, Annex B trading (AT) and an illustrative CDM case, in which Annex B parties are allowed to realize 5 per cent of the required emission reductions through CDM projects (as the difference between their Kyoto commitments and business-as-usual emissions). Change of emissions – in million tons of carbon (MtC) – as compared to the no-trade case is shown. In the trading case, non-Annex B emissions and global emissions decrease because of less leakage. The picture changes dramatically in the CDM case: due to carbon leakage associated with CDM projects, the decrease in non-Annex B does not match the increase in Annex B emissions. Global emissions increase compared to the no-trade case.

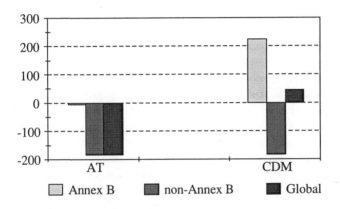

Figure 7.3   Emissions (MtC) in two scenarios: AT and CDM (relative to the no-trade case)

3   PRIVATE-SECTOR PARTICIPATION UNDER THE KYOTO MECHANISMS

3.1   Private-sector Participation in JI and CDM Projects

There is a difference between the definition of JI and the CDM concerning private-sector participation. JI Article 6 defines, among other things, the role of the national governments of Annex B parties and the potential role of legal entities in co-operation between Annex B parties on greenhouse gas abatement projects. The role of the private sector in the CDM is defined less strictly; there is no passage in Article 12 saying that credits can only be transferred to or acquired from parties. The precise role of the private sector in CDM projects is under discussion with respect to the CDM reference manual, which will be elaborated upon in Section 4.1, below. Therefore, we will now focus mainly on private-sector participation in JI.

JI Article 6.1 provides that any Annex B party 'may transfer to, or acquire from, any other such party emission reduction units resulting from projects aimed at reducing anthropogenic emissions by sources or enhancing anthropogenic removals by sinks of greenhouse gases in any sector of the economy'. In Article 6.3, the potential role of so-called legal entities is included: '[A] Party may authorize legal entities to participate, under its responsibility, in actions leading to the generation, transfer or acquisition of emission reduction units'. From this passage it follows that it is at the discretion of the government of a party whether or not to involve other legal entities in a JI project as contractors. The expression 'legal entities' is sufficiently broad to include both private-sector entities and other relevant entities such as international organizations, other public authorities and, if relevant, non-governmental organizations (NGOs). All such entities may be allowed to play a role, as contractors or otherwise, in 'actions leading to the generation, transfer or acquisition' of ERUs between states.

An example of the role of private-sector entities can be found in the ERU Procurement Tender (ERUPT) of the Netherlands government. The tender, which was launched in 2000 and completed in 2001, invited private-sector entities to propose JI projects. After approval by both the host country and the Netherlands government, the private-sector entity can start implementing the project. For the emission reductions achieved, the host country provides a claim to the private-sector entity, which it can sell to the Netherlands government. On the basis of this claim, the Netherlands government can acquire the ERUs from the host country. This demonstrates that the business sector is to play a key role in the generation of ERUs, while the actual transfer and acquisition is made by and to state parties only. The Dutch ERUPT program was extended in 2001 to include tenders for CDM projects as well (CERUPT).

3.2   Private-sector Participation under IET

The Annex B parties to the Kyoto Protocol have two basic options, which could be combined, to shape an international emissions trading system on the basis of Article 17:

- intergovernmental emissions trading (international trading of assigned amounts);
- private emissions trading (international trading of emission permits).

Intergovernmental trading would occur among governments, whereas private emissions trading would take place among private entities across national borders. The advantage of applying only intergovernmental emissions trading is that governments would remain in direct control of trading with the assigned amounts that have been defined at their level. Furthermore, it would not be necessary to allocate permits to firms. This would avoid potential legal and political problems that might arise from international differences in allocation procedures for domestic permits.

The disadvantage of intergovernmental emissions trading is that governments have incomplete information on the marginal abatement costs of domestic emitters: the higher the information deficit, the higher the probability that the enacted emissions trading deals are not as cost-effective as possible. Therefore, several economists (for example Tietenberg, 1999; Zhang and Nentjes, 1999) prefer private emissions trading to emissions trading between governments, because firms would have more and better information than governments to achieve least-cost emission trades. Governments could develop domestic permit schemes for private entities that could be connected internationally to facilitate the development of an international permit market.

Linking domestic systems internationally requires, among other things, that permits be defined uniformly (for example in tons of emissions of carbon equivalents). National agencies should, on a bilateral basis, administer the international permit trades as well as the corresponding alterations in their assigned amounts, which have to be reported to the FCCC Secretariat. Furthermore, the parties would have to meet eligibility criteria that specify an adequate structure for national compliance, such as binding national emission targets and timetables, reliable national registration and accounting of source-related emissions, accurate emissions monitoring and effective legal mechanisms for enforcement.

International differences in allocating domestic permits to private entities are not problematic from an efficiency perspective. Grandfathered permits are a lump-sum transfer, which does not alter marginal costs and therefore does not affect production decisions. Grandfathered firms or sectors do not have a cost advantage over identical firms or sectors abroad which are auctioned, because the former have to include the opportunity costs of using the permits in the product price. Efficient competition is not distorted. However, such differences could be interpreted as state aid under EC (European Community) law and/or actionable subsidies under WTO (World Trade Organization) law if the equity aspect of permit allocation is considered to be relevant (Woerdman, 2000). Grandfathered permits are a capital gift to the firm, inducing a windfall profit, which implies that a similar firm abroad that has to buy its permits has a higher cash outflow and hence fewer financial resources. Although these financial consequences have no effect on overall efficiency, they could be considered as inequitable. Grandfathering could be perceived as 'distorting' in the sense that it gives an unfair advantage to the grandfathered firm compared to an auctioned firm and changes the 'level playing field' between them. Fair competition is then distorted. Whether it will actually be interpreted in this way remains to be seen. There are several political precedents, such as tradable fishing quotas, that suggest that allocation differences are acceptable from an efficiency point of view.

The reasoning behind not using the information advantage of firms – because a different and more general aspect of the Kyoto Protocol (namely compliance) is not well organized – is highly questionable. Rather, governments should first establish compliance procedures at the national level. Such a scheme would create transparency regarding the party's performance in complying with its international emission commitment. If a party did not comply, it would be possible to identify and impose sanctions on the private sources that emit more than their permits allow. This enhances the effectiveness of the international environmental agreement. Furthermore, the direct participation of firms in the international greenhouse gas emissions trading system is expected to create a 'thick' market with many small traders, which avoids market power and enhances the efficiency of climate policy.

4   ENVIRONMENTAL INTEGRITY OF THE KYOTO MECHANISMS

4.1   Baseline Determination in JI and CDM Projects

In a JI or CDM project, emission reductions are measured bottom-up from a baseline, which estimates what future emissions at the project site would have been if the project had not taken place. The emission reduction is calculated as the difference between these baseline emissions and the (lower) emissions measured during the project's crediting lifetime. This emission reduction must be in addition to what would have occurred otherwise. If future emissions are overestimated by setting the baseline too high, emission reductions that have not in fact occurred are credited.

In general, the main problem with respect to JI and CDM baselines is to find a methodology that balances maximizing the environmental integrity of a project and minimizing the transaction costs. For example, trying to maximize environmental integrity by collecting as much information as possible about a project's reference case may reduce the project's cost effectiveness. Setting up lenient procedures for baselines in order to minimize transaction costs may result in certified emission reductions that in reality do not exist. Although a strict third-party verification procedure could reduce the latter effect, it would increase transaction costs (which would probably be shifted to the project developers). This implies that baseline determination still incorporates an unavoidable trade-off between cost reduction and environmental integrity. Standardizing procedures for baseline determination is an option that could soften this trade-off. Baseline standardization primarily aims at lowering the transaction costs of baseline determination, while at the same time maintaining the project's environmental integrity at an acceptable level. Standardization implies the development of 'business-as-usual' scenarios for project categories differentiated by, for instance, region, time, project and/or technology type and determined by a panel of experts. Each specific project should fit into one of the categories. The (*ex ante* or *ex post*) emission reductions of a specific project can simply be calculated by subtracting the (predicted or observed) emissions from the baseline emissions of the relevant category.

A practical example of standardization is the use of benchmarks, which define standard emission factors for a certain project in a particular host country. These factors could be derived, for instance, from default projects (Luhmann et al., 1995), recent historical country/sector/fuel averages or recent/future marginal technology as proxies of the most likely investment. Benchmarks can be presented in a matrix where standard emission factors are listed per sector/project type and country/region (for example Jepma et al., 1998). A potential problem with benchmarks is that free riders may enter JI and the CDM. If a benchmark is set for a particular host country at a certain baseline level, each company that beats this benchmark can apply for credits, including those that would have implemented the project anyway for commercial reasons. This raises the question as to what extent baseline standardization sufficiently deals with ensuring additionality. Nevertheless, there are several options to accommodate some of these concerns, such as limiting crediting lifetime, determining conservative benchmarks and/or defining default additionality standards next to default baselines.

It could be argued that baseline determination is less problematic for JI than for the CDM. The host country of a JI project has committed itself to an assigned amount, which implies that the government has to define environmental policy targets for its domestic emitters. If it has done so, the JI baseline to calculate the additional emission reductions could simply be derived from the defined environmental policy for the host firm or sector involved. For instance, if the policy in the host country is a relative (or *performance*) standard that requires a certain quantity of CO_2 per unit of output or energy, the baseline (or *benchmark*) emissions for the host firm can be calculated by multiplying this standard with its expected production volume. If the host firm emits less CO_2 than this baseline figure because a certain abatement project has been implemented, emission reductions are achieved for which the investor can obtain credits.

However, there is still the risk of a 'micro-macro mismatch' for JI baseline determination. Although the host country's emission target under the Protocol allows a kind of top-down baseline approach, it does not remove the incentive for project developers at the micro-level to inflate the baseline in order to claim more credits. Consequently, this would require careful verification procedures to prevent the host country from transferring too many credits to investors and running the risk of being in non-compliance. The option of baseline standardization could assist JI host countries in such a procedure.

A more general issue related to JI and CDM baseline determination is whether the whole project cycle (including baseline determination, accreditation, monitoring and verification) for JI and CDM projects should be similar. One option is to apply a CDM-type project cycle for JI to ensure environmental integrity and to let JI projects be validated and certified by an independent operational entity (possibly supervised by the Executive Board under the CDM). Another option is to construct a less stringent project cycle for JI. This could mean that Annex B parties could determine a JI project baseline themselves (for example following national guidelines) and start a project, provided that both the host and the investor are in compliance with Articles 5 and 7 of the Kyoto Protocol dealing with national greenhouse gas inventory systems, communication and information. A so-called 'fast' track for JI has in fact been created at COP7 in 2001. If a host party has a reliable emission registration/inventory system (according to the aforementioned Protocol articles), a host party may verify reductions as being additional. Otherwise, verification must occur along a 'slow' track via the Article 6 Supervisory Committee more or less similar to the CDM project cycle.

An advantage of a less stringent project cycle for JI is that transaction costs would be relatively low. However, a less stringent procedure for JI projects might imply a competitive disadvantage for CDM projects in developing countries. Furthermore, it might increase the risk of a 'micro-macro mismatch' under JI, as explained above. In addition, a more practical disadvantage of a less stringent project cycle for JI is that compliance with the reporting and inventory commitments could only be determined one year before the commitment period at the earliest. This could, given the preparation time usually required for JI projects, create a disincentive for early JI action. One policy option, therefore, is to develop international criteria for JI project baselines and for verification and certification procedures. Based on such criteria, accredited independent certifying entities could check project baselines as well as verify and certify the projects' results. A positive judgement by a certifying entity would then give the investor and host Parties the 'go-ahead' with the JI project irrespective of their compliance with Article 5 and 7.

There seems to be a majority view among parties that the CDM project cycle (including baseline validation, monitoring and verification) needs to be formalized through a reference manual (for example, SBSTA/SBI, 2000). The technical details for the CDM baseline methodology to be included in the reference manual will probably be decided at meetings of the Conference of the Parties (COP) in the near future (possibly during 2003). However, current negotiations suggest that during the crediting period of a CDM project, a validated baseline methodology would not be subject to revision unless recommended by an independent verification entity. Furthermore, there seems to be considerable agreement among the FCCC Parties that a revision of the methodology would only be relevant to future crediting periods and would not affect existing projects during the current crediting period.

At COP6 Part II in Bonn, the parties agreed that afforestation and reforestation projects are eligible under the CDM. However, the total of subtractions from and additions to a party's assigned amount resulting from these CDM projects shall not exceed 1 per cent of its base-year emissions. Forest preservation projects are excluded from the CDM, at least for the first commitment period, because these are considered to be more uncertain in terms of technical methodologies for adequately calculating and monitoring carbon sequestration in existing forests (CP, 2001). Such forest management is allowed in JI host countries, which have an emission ceiling, but its use is restricted by quota set for each individual country. The parties also decided to refrain from the use of nuclear energy under JI and the CDM.

4.2 Hot Air under IET

Because of their negotiating power and mutual economic interests, in combination with a lack of negotiation time and incomplete information at COP3 in 1997 in Kyoto, the assigned amounts for some parties have been allocated at higher levels than their actual and/or projected future macro-baseline emissions. According to several studies, these parties – notably the Russian Federation and the Ukraine – will be able to transfer this surplus of so-called 'hot air' to other Annex B countries by means of IET. This would lead to the trading of emission reductions on paper, which have in fact not taken place. In a survey by Zhang (2000) of several different models, the hot-air projections vary between 92 and 374 MtC-eq., roughly somewhere between zero and 50 per cent of the required Kyoto efforts, depending on the estimated level of business-as-usual emissions. However, Bashmakov (1999) seems to consider the tradable hot air in Eastern Europe as legitimate compensation for the emission reductions induced by the economic decline that resulted from the disintegration of the centrally planned economic system.

Hot air is perceived by many parties, scientists and NGOs as an important political problem. Still, it could be argued from a dynamic perspective that countries such as the USA (who abandoned the Protocol in March 2001), Japan and the Russian Federation would not have accepted commitments at COP3 if expectedly cheap hot air had not been available to buy (in the case of the USA and Japan) or sell (in the case of Russia). This dynamic perspective is underlined in a simple rational-choice model by Boom (2000), who demonstrates that these Annex B countries would only have committed themselves to a much less stringent emission ceiling (or even no ceiling) had the EU blocked an agreement with hot air in Eastern Europe. In particular, such an increase in the overall emission cap might have exceeded the volume of hot air in the Kyoto Protocol.

On the one hand, this dynamic perspective suggests that limiting the availability of hot air could lower the chance of acceptance and ratification of the Protocol by the Annex B countries and could make them unwilling to engage in future agreements on emission reductions. On the other hand, hot air remains problematic from a static perspective, because, with emissions trading, the hot air can be sold during the first commitment period and overall emissions will then be higher than they would have been without emissions trading. There are several possible options for dealing with the hot-air issue. We will discuss four of them.

First, prior to COP6, the EU (supported by countries such as China and India) proposed limiting hot air by placing a quantitative restriction on the use of the Kyoto mechanisms (SBSTA/SBI, 2000). However, the EU gave up its proposal as part of a compromise reached at COP6 Part II in Bonn (see Section 5 for more details). The restriction on trade, as initially proposed by the EU, would not be effective: it would probably limit hot air to one-third of its potential magnitude (Baron et al., 1999). Furthermore, even if it were effective, the hot air countries, such as the Russian Federation, would still have the opportunity to use hot air in the future by banking it to a subsequent commitment period on the basis of Article 3.13. In addition, a ceiling on trade would be likely to increase carbon leakage to non-Annex B countries. Trade restrictions are inefficient and increase the costs of greenhouse gas control, thus raising the price of energy-intensive products in Annex B countries, which gives developing countries a comparative advantage in those markets. As a result, the developing countries, which would not be restricted by formal targets, would increase energy-intensive production, and global emissions would rise, thereby mitigating the reduction of hot air.

Second, another alternative to dealing with the hot-air issue was proposed by Switzerland in the run-up to COP6: the only units that would be eligible for transfer via IET would be those that are clearly backed up by greenhouse gas emissions reductions beyond business-as-usual emissions. For example, suppose a party is allowed 100 units of greenhouse gas emissions per year during the period 2008 to 2012. According to its business-as-usual emissions scenario, the emissions turn out to be only 90 units per year. Under this proposal, the party could only transfer assigned units via IET if it reduced its greenhouse gas emissions to a level lower than 90 units per year. The 10 units of business-as-usual emission reductions would, under this proposal, not be eligible for trading. The disadvantage of the Swiss proposal is that it remains rather difficult to determine the business-as-usual scenario for a country. In addition, if a system of 'excess emissions reductions' were applied through IET, the amount of excess reductions would have to be known beforehand in order not to frustrate the scope for early IET transfers. However, this appears to be difficult as long as there are no clear, internationally agreed-upon rules about who will determine the amount of excess reductions, and how.

Third, rough guidelines for another possible compromise have been sketched by Nentjes and Woerdman (2000), who propose that the EU should buy, set aside and never use the hot air, while at the same time accepting unrestricted trading. This combination would have several advantages: the EU would show leadership and get an effective protocol without hot air. The Russian Federation would earn money for its hot air, and industrialized countries, such as the USA and Japan, would be able to trade efficiently without quantitative restrictions. The EU would show responsibility without affecting developing countries: the elimination of hot air would imply stricter reduction targets for industrialized countries, and the developing countries would not be restricted from receiving money and technology via CDM projects. However, this proposal has the same practical problems as the Swiss proposal, given the potential difficulty in predicting the available amount of hot air. Moreover, it is politically difficult to decide how to provide the money for the acquisition and retirement of hot air, although there are some options available. For instance, it could be partly financed from the revenues of a transaction tax on all Kyoto mechanisms (similar to the current adaptation tax defined only for CDM transactions in Article 12.8), which would be less restrictive than a quantitative ceiling on trade.

Fourth, a more indirect option to exclude hot air while maintaining an effective and efficient (albeit potentially smaller) trading system would be to demand that participants satisfy certain eligibility criteria, with respect to accurate monitoring and enforcement, for instance, before they are allowed to trade. Many observers believe that this should be advocated irrespective of the hot-air problem in order to obtain a credible trading system. However, the result could be that the eligibility requirements de facto exclude those parties that are also likely to supply the hot-air quota, because the Russian Federation and Ukraine may be unlikely to meet the requirements in the short term (before and/or during the first commitment period). One of the disadvantages of this eligibility approach is that it potentially excludes those Annex B countries that also have the cheapest abatement options from the trading system, thereby reducing the cost-saving potential of the Kyoto mechanisms. Moreover, the possible exclusion of notably Eastern European countries would deprive them, in principle, of the option to reduce the costs of meeting their Kyoto commitments by means of emissions trading, which could be perceived as unfair.

It is clear that each of these options for eliminating hot air contains one or more problems to be solved or trade-offs to be made – between efficiency and equity, for instance, or between effectiveness and acceptability. Moreover, although hot air is not so much a problem from a dynamic perspective, it is perceived as a political problem by various actors with different interests, perceptions, values and negotiating power. Excluding the hot air from trading is unlikely to be acceptable for potential buyers, such as Japan, and for potential sellers, such as the Russian Federation, whereas leaving the hot air untouched is undesirable for those who stress the environmental integrity of the commitments of industrialized parties, such as the EU and the environmental NGOs. Nevertheless, the withdrawal of the USA from the Kyoto Protocol in March 2001 changed the game and seems to have increased the acceptance of hot air by the EU and the NGO community as a 'necessary evil' to keep other Annex B parties, such as Japan and the Russian Federation, on board.

5   RESTRICTIONS ON USING THE KYOTO MECHANISMS

5.1   Reasons for a Trade Restriction

The Kyoto Protocol indicates that use of the Kyoto mechanisms shall be supplemental to domestic action (Articles 6.5c, 12.3b and 17). This controversial issue was one of the main reasons that no agreement was reached at COP6 Part I in 2000 in The Hague. Countries gathered in the so-called 'umbrella group', such as the USA, the Russian Federation and Japan, did not want to elaborate the term 'supplementarity'. They stressed that domestic action is only supplemented by the Kyoto mechanisms if domestic marginal abatement costs are higher than the permit price. However, the EU, supported by countries like China and India, wanted to elaborate supplementarity by placing a quantitative restriction (or *ceiling*) on the use of the Kyoto mechanisms (SBSTA/SBI, 2000). Roughly, the EU wanted to limit trade by 50 per cent, and its proposal contained specific restrictions on both import (demand) and export (supply).

In an empirical analysis based on a content analysis of official EU documents and on interviews with key EU decision-makers, Woerdman (2002) found that such a ceiling on trade was mainly advocated on the basis of three considerations: namely, hot air, equity and compliance. Proponents of a ceiling on trade argued that it would reduce hot air because it reduces emissions trading. They also believed that limited use of the Kyoto mechanisms would be fair, since it would prevent Annex B countries from 'buying their way out' cheaply. In addition, they feared that trading would increase the risk of non-compliance, because countries might sell too many emission rights.

To some extent, the EU seemed to be internally divided on the issue of supplementarity, although the EU ministers kept their ranks closed at COP6. Just before COP6 Part I, a number of officials and decision-makers from almost all EU member states were interviewed, and speaking in a personal capacity, almost half expressed their disagreement with the EU proposal of a ceiling on the Kyoto mechanisms (Woerdman, 2002).

5.2 Consequences of a Trade Restriction

Nentjes and Woerdman (2000) argue that trade restrictions, as initially proposed by the EU, would be inefficient and would endanger compliance. A quantitative ceiling on all Kyoto mechanisms would raise the cost of reducing emissions, which would make compliance more difficult. Moreover, such a ceiling on trade is likely to be ineffective: it would probably limit hot air to only about one-third of its potential magnitude and it could increase carbon leakage. In addition, restricting trade can also be seen as inequitable: it denies developing countries a potential source of additional income by also limiting the CDM. Finally, it is wrong to think that domestic action disappears when there are no restrictions on trade: even in a full-trade scenario, Annex B would still implement about 30 to 50 per cent of its required reduction efforts domestically according to the models of Zhang (2000) and Bollen et al. (2000), respectively.

While, on average, net acquiring countries will face higher costs as a result of quantitative restrictions on the Kyoto mechanisms, compared to the unrestricted case, individual importing countries could actually gain (Bollen et al., 1999). This would be the case with a country that is not affected by the restriction because it imports only a limited amount. The lower price for transferred emission reductions implies lower import costs. It even means that this unrestricted importing country could now import more than it could under fully competitive trade. This would lead to less domestic abatement (that is the ceiling would have a perverse impact) and a decrease in the burden. Even when a country is slightly restricted, ceilings could lower its burden because the lower import costs would outweigh the higher domestic abatement costs. But more restricted countries would experience a serious increase in the burden. A ceiling on the use of the Kyoto mechanisms could thus have an asymmetric impact on the burden of countries.

On the basis of the WorldScan model, most OECD importers are seen to be constrained by the restriction and to make a greater effort at domestic reduction. For instance, the restriction harms the EU, whose domestic marginal abatement costs are considered to be relatively high. (The cost increase outweighs the advantage of lower prices on imported permits.) However, the restriction is not binding for the USA, whose domestic marginal abatement costs are assumed to be relatively low. Because the price of transferred entitlements would decline, the USA would increase its imports and take less domestic action (Bollen et al., 2000). This means that the EU would lose if its own proposal to restrict trade had been adopted.

Compared to restrictions on importing emission permits, restrictions on the export of emission permits would shift the burden, with the result that importing countries would lose unambiguously, since the permit price would go up. For exporting countries, the result is less clear. The higher permit price might compensate for the lower export volume. Restrictions on both import and export could be a way to lower the sale of hot air.

Again, the withdrawal of the USA from the Protocol changed the game: at COP6 Part II in Bonn, the EU was willing to give up its proposal and accept the unspecified requirement that domestic action should be a 'significant element' of Annex B countries' climate policy (CP, 2001). The EU made this compromise to prevent the other umbrella countries from following the USA and abandoning the Protocol as well. Moreover, some of the aforementioned environmental concerns of the EU were accommodated by means of restrictions on the use of sinks (as discussed in Section 4.1) and the requirement, among other things, that each Annex B party maintains a reserve during the commitment period that should not drop below 90 per cent of its assigned amount or 100 per cent of five times its most recently reviewed inventory, whichever is lowest. The implication of the latter requirement is twofold. On the one hand, a hot-air country (in fact any country) is allowed to sell 10 per cent of its hot air without any repercussions. On the other hand, a hot-air country is allowed to trade more hot air only if and to the extent that its reviewed emissions are lower than 90 per cent of its assigned amount.

6   CONCLUSIONS

The Kyoto mechanisms lower the costs of reducing emissions, but several political, environmental and institutional barriers complicate their implementation, such as those related to private-sector participation, project baseline determination and hot-air trading.

The private sector is already allowed to play a role in the development as well as the validation, monitoring and certification of JI and CDM projects. Private-sector participation is also essential for international emissions trading to obtain a 'thick' and efficient market without dominant market power. If domestic permit trading systems are developed, they can be connected internationally in the long term, provided that countries have accurate monitoring and strong enforcement structures.

Standardizing procedures to determine baselines for JI and CDM projects is a potential approach to finding a balance between securing environmental integrity and reducing transaction costs. However, standard baselines carry the risk of free riders that could possibly acquire JI/CDM credits for projects that are not additional. This may require additionality checks.

Looking back on the negotiation process on climate change, it seems possible that, without hot air, several potential buyers (such as Japan) and sellers (such as the Russian Federation) might only have accepted less stringent commitments or even none at all. Looking forward to the approaching commitment period(s), hot air may lead to the trading of entitlements that do not represent genuine emission reductions. To limit hot air, the EU initially proposed a quantitative restriction on the use of the Kyoto mechanisms. Although its underlying concern for environmental integrity was sincere, such a restriction would have raised costs and would have limited income transfers to developing countries. The withdrawal of the USA from the Protocol changed the game and induced the EU to give up its proposal to prevent other Annex B countries from abandoning the Protocol as well.

These and other findings discussed in this chapter allow us not only to characterize the first commitment period of 2008 to 2012, but also to provide an evolutionary perspective on a possible second commitment period. For the first commitment period, the Kyoto mechanisms are available for industrialized countries to lower the costs of reducing emissions, with restrictions on the use of sinks. Developing countries can attract income and technology for sustainable development via the CDM. Industrialized countries and countries with economies in transition may use JI, and international emissions trading between firms could be made available for eligible countries that have domestic tradable permit schemes with strong monitoring and enforcement structures.

For a second commitment period, the scope for carbon leakage would be reduced by the policy option in which more countries adopt targets for reducing emissions. If the least poor and most polluting non-Annex B countries would be willing to accept binding (growth) targets in the future, they could host JI instead of CDM projects. Moreover, previously ineligible Annex B countries that have improved their monitoring and enforcement over time could be allowed to let their firms engage in permit trading across national borders, if desirable. In the long term, this would facilitate the development of an efficient and effective international market for carbon trading with sound private-sector participation.

NOTE

1. See for example Capros et al. (2000) who compare ten models (such as POLES, G-CUBED and EPPA).

REFERENCES

Banuri, T. et al. (2001), *Climate Change 2001: Mitigation*, Technical Summary, Third Assessment Report, Geneva: Intergovernmental Panel on Climate Change (IPCC).

Baron, R., M. Bosi, A. Lanza and J. Pershing (1999), *A Preliminary Analysis of the EU Proposals on the Kyoto Mechanisms*, Draft, 28 May/8 June, Paris: International Energy Agency (IEA).

Bashmakov, I. (1999), 'Strengthening the Economy through Climate Change Policies: The Case of the Russian Federation', in C.J. Jepma and W.P. van der Gaast (eds), *On the Compatibility of Flexible Instruments*, Dordrecht: Kluwer, pp. 17–30.

Bollen, J.C., A.M. Gielen and H.R. Timmer (1999), 'Clubs, Ceilings and CDM: Macroeconomics of Compliance with the Kyoto Protocol: The Costs of the Kyoto Protocol: A Multi-model Evaluation', *Energy Journal* (Special Issue), 177–206.

Bollen, J.C., A.J.G. Manders and P. Tang (2000), *Winners and Losers of Kyoto*, The Hague: Dutch Central Planning Bureau (CPB).

Boom, J.T. (2000), *The Effect of Emission Trade on International Environmental Agreements*, Working Paper 00-1, Aarhus, Denmark: The Aarhus School of Business, Department of Economics.

Capros, P., L. Mantzos, M. Vainio and P. Zapfel (2000), 'Economic Efficiency of Cross Sectoral Emission Trading in CO_2 in the European Union', Paper presented at the Conference on Instruments for Climate Policy: Limited versus Unlimited Flexibility? 19–20 October, 2000, Ghent, Belgium: University of Ghent.

CP (2001), *Implementation of the Buenos Aires Plan of Action.* Decision 5/CP.6 ('Bonn Agreement'), Document FCCC/CP/2001/L.7, 24 July 2001. Bonn: United Nations Framework Convention on Climate Change (UNFCC).

CPB (1999), *WorldScan: The Core Version*, The Hague: Dutch Central Planning Bureau (CPB).

Jepma, C.J., W.P. van der Gaast and E. Woerdman (1998), *The Compatibility of Flexible Instruments Under the Kyoto Protocol*, NRP Report No. 410 200 026, Bilthoven, Netherlands: Dutch National Research Programme on Global Air Pollution and Climate Change (NRP).

Luhmann, H.-J., C. Beuerman, M. Fischedick and H. Ott (1995), *Making Joint Implementation Operational: Solutions for Some Technical and Operational Problems of JI in the Fossil Fuel Power Sector*, Wuppertal Papers No. 31, Wuppertal, Germany: Wuppertal Institute.

Michaelowa, A. and M. Dutschke (1998), 'Interest Groups and Efficient Design of the Clean Development Mechanism under the Kyoto Protocol', *International Journal of Sustainable Development*, **1**(1), 24–42.

Nentjes, A. and E. Woerdman (2000), *The EU Proposal on Supplementarity in International Climate Change Negotiations: Assessment and Alternative*, ECOF Research Memorandum 2000 (28), Groningen, Netherlands: University of Groningen (RUG).

SBSTA/SBI (2000), *Mechanisms Pursuant to Articles 6, 12 and 17 of the Kyoto Protocol*, text for further negotiation on principles, modalities, rules and guidelines. SBSTA/SBI Twelfth Session, FCCC/SB/2000/3, Bonn, 12–16 June, 2000, Bonn: United Nations Framework Convention on Climate Change (UNFCC).

Tietenberg, T. (1999), 'Lessons from Using Transferable Permits to Control

Air Pollution in the United States', in J.C.J.M. van den Bergh (ed.), *Handbook of Environmental and Resource Economics*, Cheltenham, UK and Northampton, MA, USA: Edward Elgar, pp. 275–92.

Woerdman, E. (2000), 'Competitive Distortions in an International Emissions Trading Market', *Mitigation and Adaptation Strategies for Global Change*, 5(4), 337–60.

Woerdman, E. (2002), 'Why Did the EU Propose to Limit Emissions Trading? A Theoretical and Empirical Analysis', in J. Albrecht (ed.), *Instruments for Climate Policy: Limited Versus Unlimited Flexibility*, Cheltenham, UK and Northampton, MA, USA: Edward Elgar.

Zhang, Z.X. (2000), 'Estimating the Size of the Potential Market for the Kyoto Flexibility Mechanisms', *Weltwirtschaftliches Archiv (Review of World Economics)*, 136(3), 491–21.

Zhang, Z.X. and A. Nentjes (1999), 'International Tradable Carbon Permits as a Strong Form of Joint Implementation', in J. Skea and S. Sorrell (eds), *Pollution for Sale: Emissions Trading and Joint Implementation*, Cheltenham, UK and Northampton, MA, USA: Edward Elgar, pp. 322–42.

8. Terrestrial carbon sinks and biomass in international climate policies

Jelle G. van Minnen, Ekko C. van Ierland and Gert-Jan Nabuurs

1   INTRODUCTION AND BACKGROUND

Recent changes in climate are at least partly due to emissions of CO_2 and other greenhouse gases from human activities (Mitchell et al., 2001). To limit climate change and its potential impact, countries met in Kyoto in 1997 to supplement the United Nations Framework Convention of Climate Change (UNFCCC) with quantitative targets for industrial countries (the so-called 'Annex B') to limit their greenhouse gas emissions. In addition to reducing emissions from fossil fuel burning, the Protocol makes provisions for Annex B countries to achieve parts of their reduction commitments by taking into account various land-use options. These options may include direct action within the land-use and forestry sectors, such as afforestation, reforestation and deforestation (ARD, Art. 3.3), and other activities related to land-use (Art. 3.4), such as forest management and improved management of land, such as reduced tillage, and changes in land-use, such as more grassland instead of crop land).

After the Kyoto Protocol was signed, more Conferences of Parties (COP) of the FCCC were held to resolve outstanding issues that are critical for ratification of the Protocol. One main outcome of the recent COPs (in Bonn and Marrakech) is an agreement on the way of dealing with these land-use, land-use change and forestry (LULUCF) measures (see details in Section 6, below).

The objective of this chapter is to discuss some issues related to the biophysical potential of different land-use options, as well as the economic and political implications. Our aim is to discuss these issues in a policy context. Literature referenced in relation to the terrestrial biosphere and global carbon cycle includes Goudriaan (1995), Van Minnen et al. (1995), Schlamadinger and Marland (1998) and Noble and Scholes (2000). In Section 2, we first identify why the terrestrial carbon sinks and biomass were politicized, and why they have become controversial. Second, we briefly quantify the potential of using C sinks and biomass as mitigation options (Section 3). In Sections 4 and 5, we present the economic issues of using sinks and biomass systems, showing, for example, the economic characteristics of land-use projects under the clean development mechanism (CDM). In Section 6, political views are discussed with respect to the inclusion or exclusion of land-use options in mitigating climate change. The section includes a summary of the achievements made at the recent Conferences of Parties in Bonn. Finally, in Section 7, we summarize remaining issues and present conclusions about the capacity of using C sinks and biomass systems as mitigation options.

2　WHY IS THE C SINK ISSUE SO CONTROVERSIAL? OPEN QUESTIONS, REMAINING ISSUES AND DEFINITIONS

The 1997 Kyoto Protocol represents a milestone in the negotiations on climate policy with binding targets being established. Especially with respect to sinks and biomass, the Protocol is not always explicit and contains issues that need to be defined or clarified for ratification (Schlamadinger and Marland, 2000). This has led to considerably different views among countries (Metz et al., 2001). Political progress has been made since 1997 in resolving these issues, especially during the meeting in Bonn in July 2001 (COP 6 bis) (see Section 6, below). However, some open issues remain, such as the permanence of sink projects in non-Annex B countries (see also Section 6). In addition, there are several reasons why the terrestrial carbon sinks have become so politicized and controversial:

- Even conservative estimates of the sink and biomass capacity show a considerable potential to meet the Kyoto commitments. Some countries fear that the use of sinks as a mitigation option might reduce the incentive in some countries to reduce their fossil fuel emissions.
- Large countries with large forests and agricultural areas were thought to be able to meet their reduction target easily by using sinks alone. In some cases, they would even be able to increase their fossil-fuel emissions. It was felt that this did not equally distribute the burden of reducing emissions.
- The accounting, prevention of leakage, and so on could lead to a huge global bureaucracy, keeping track of all flows of carbon in each little parcel of land.

- Different methods have been proposed (see Section 3, below) to quantify the sink and biomass potential. These methods give different outcomes, resulting in uncertainties and opportunities for policymakers.
- The terrestrial carbon cycle is inherently dynamic and only partly influenced by human activities. Other contributing factors include, for instance, variations in weather, climate and natural disturbances. It is within this uncertain framework that binding targets have to be discussed.

3 QUANTIFICATION OF C SINKS AND BIOMASS USE

Terrestrial sinks and biomass are considered very important for the global carbon cycle and may thus contribute to solving the climate problem. One way of categorizing related land-use activities is by grouping them according to the mechanisms through which they affect carbon stocks and flows and, thus, the CO_2 concentration in the atmosphere. In principle, there are three basic mechanisms (Kauppi et al., 2001):

- carbon sequestration (that is sequestering carbon in ecosystems (including soils) and products);
- carbon conservation (that is avoiding emissions related to land use by protecting existing carbon pools (for example through forest management));
- carbon substitution (that is the replacement of fossil fuels by biomass energy).

3.1 Accounting Rules, Indicators and Measurement Methods

An accounting system is needed to record and report changes in carbon stocks from applicable land-use activities. According to the Kyoto Protocol, this system, which has to be defined by 2005, should be transparent, consistent, verifiable and relatively accurate in quantifying the potential C sink for Annex B countries during the first commitment period. But what should such an accounting system look like? Various factors have a strong influence on the sink potential of an activity. Matthews et al. (1996) and Nabuurs et al. (1999), for example, showed that the methods and criteria used within an accounting scheme are important for determining the net potential of sinks and biomass as mitigation options. There are two reasons for this.

First, two different accounting approaches have been discussed within the climate arena that could meet the requirements of transparency, consistency and verifiability. These are (1) a *land-based* and (2) an *activity-based* accounting scheme (Watson et al., 2000). Both schemes have both advantages and disadvantages (Box 8.1).

BOX 8.1   ACCOUNTING METHODS

The basis of the 'land-based' accounting scheme is that, first, applicable activities are defined, followed by an identification of 'Kyoto land' where these activities may occur. This approach involves full carbon stock changes on this land between 2008 and 2012. Adjustments can easily be made on baselines, leakage, and so on. A disadvantage of a land-based accounting system is that it can be difficult to separate human-induced stock changes from indirect effects. Further, fluxes in non-CO_2 gases cannot be estimated (Noble and Scholes, 2000).

The 'activity-based' approach starts with defining certain activities, followed by accounting for changes in C stocks for each of the applicable activities (per unit area and time). The stock changes are then multiplied by the area on which an activity occurs and the number of years the activity is applied. A disadvantage of the 'activity-based' approach is that an area could potentially be counted more than once if multiple activities occur and the activities are not supplementary, which could result in inaccurate accounting. Alternatively, each land unit could be subject to only one activity.

Second, the question is which stocks or fluxes should be considered and how should they be compared in 2010 to those considered in 1990 (Box 8.2). This is important, because the land-use change and forestry (LUCF) sectors include, contrary to other sectors, sinks and emissions of carbon. Should only the sinks be considered or should it be the net C flux? Different proposals have been made (see Noble and Scholes, 2000, for details). The Kyoto Protocol specifies what is often referred to as the 'gross–net' approach. In this approach, assigned amounts are based on gross emissions in 1990 (not corrected for sinks), but emissions in the first commitment period are based on net emissions (that is countries can get benefits for their sinks). The problem with the 'gross–net' approach is the large uncertainty of sinks, which would only count in setting the target. The sinks in the first commitment period may be very large, especially if the approximately 2.3 Gt C.yr^{-1} of net carbon sequestration in the biosphere is identified as especially

BOX 8.2 INDICATORS TO QUANTIFY THE SINK POTENTIAL

In general, the sink potential of an area or activity can be quantified by measuring either changes in C stocks or C fluxes. An advantage of considering changes in fluxes is that fluctuations in carbon sequestration or release between years can be considered. This may be important for determining the change during the different commitment periods. A disadvantage of the flux measurement is that it only measures the dynamics over a relatively short time period.

For the *stock approach* – deeply rooted in the field of forestry and agriculture (for example forest inventories) – one can consider changes in above-ground biomass, possibly expanded with wood products. This approach is highly verifiable; however, the method is weak at accounting for large amounts of carbon stored in different soil compartments. This could be solved by considering the *soil carbon pools* and *soil processes* as well, although it is complicated by the lack of accurate soil data (which is important in determining the baseline) and the difficulty in verifying small changes in these data.

There is also a variety of indicators to measure fluxes. One can consider the net primary production NPP) or total growth of an ecosystem. NPP is the CO_2 taken up through photosynthesis minus the CO_2 loss due to plant respiration. Similar to the approach mentioned above that measures above-ground biomass, the NPP approach is highly verifiable, but it does not take soil processes and pools into consideration. If changes in pools of soil carbon become part of the accounting scheme, net ecosystem productivity (NEP) would be an alternative measure. NEP describes the net carbon uptake of an ecosystem (including soils), assuming no changes in land use or cover. A disadvantage of using NEP is that it is difficult to verify. This problem is even larger for net biome productivity (NBP), which denotes an ecosystem's net carbon uptake/release. Compared to the NPP and NEP fluxes, NBP is relatively small (about 1 and 10 per cent, of NPP and NEP, respectively). The advantage of NBP is that, compared to the NEP approach, it also considers other processes that lead to carbon loss (for example harvests and natural disturbances). In regard to the Kyoto Protocol, NBP still does not account for the fate of the original vegetation on that site and its carbon budget. This could, however, be important if someone wanted to quantify the net implications of ARD activities. The surplus potential productivity (SPP) has been developed in an attempt to account for the 'real' carbon gain of planting a Kyoto plantation (Onigkeit et al., 2000). SPP is defined as the NEP of a growing Kyoto forest, corrected for the carbon uptake that would have occurred if the original vegetation had remained at that location and the release that results from clearing the original vegetation for planting the forest. A positive SPP value indicates that it is worth planting a Kyoto plantation, since it will take up more carbon from the atmosphere over time than the original vegetation.

occurring in Annex B countries.

At the COP 6bis in Bonn, parties agreed on applying the so-called 'net–net' accounting approach for activities under Art. 3.4 (see Section 6, below). This approach is based on comparing the net emissions in the first commitment period with the net emissions in 1990. Thus, terrestrial sinks are considered in both the cases as the target. Some parties recognise that this 'net–net' approach would make it difficult for countries with a large sink in the base year (1990) to maintain (or even increase) the size of the sink into the commitment period.

Regardless of the accounting system that will become accepted, the potential sink should be 'reported in a transparent and verifiable manner'. Several methods have been developed to quantify changes in terrestrial carbon stock and fluxes (for example Nabuurs et al., 1999; Sarmiento and Wofsy, 1999; Dolman et al., 2001; see Box 8.3).

These methods include constraints from atmospheric chemistry (for example Fan et al., 1998), various models (for example Schimel et al., 2000), land-use bookkeeping (for example Houghton et al., 1999), flux towers (Baldocchi et al., 1996; Valentini et al., 2000) and forest inventories (Nabuurs et al., 1998; UN-ECE/FAO, 2000). What these methods all have in common is that they show a highly dynamic terrestrial biosphere. In addition, they come up with two factors that determine the carbon-sink potential in an area: the type and condition of the ecosystem (that is species composition, age and structure, site conditions like soils and climate, and management) and the (previously mentioned) question of whether changes in C stocks or C fluxes should be considered for quantifying the sink potential (Box 8.2). Important differences between the methods are that they operate at different spatial and temporal scales, include different C stocks and describe different processes. Thus, full accounting of the carbon balance inevitably requires a multi-method exercise (Nabuurs et al., 1999; Watson et al., 2000).

BOX 8.3   MEASUREMENT METHODS

Various techniques have been developed to quantify terrestrial carbon sources and sinks. All of them have both advantages and disadvantages.

Inventories were designed to sample the status of forest resources across large regions. A variety of assumptions are required to convert inventory measurements to carbon budgets. Their main limitation is related to the effort required to make them, which makes frequently repeated inventories relatively rare outside Annex B countries. Further, inventory data are often difficult to compare between countries because of, among others, differences in definitions.

The strength of measurements (for example through *eddy flux towers*) is the integrated signal from all of the mechanisms affecting the net carbon production in the ecosystem, the ability to measure gas fluxes directly and the stability of the system. But these measurements are local and limited to a few sites and thus do not capture the variability of carbon flux conditions across the broader landscape. In addition, flux tower measurements have a temporal disadvantage. The towers may measure a C sink for a number of years but may fail to measure a disturbance.

Large-scale measurements (for example *though air sampling)* and *remote sensing.* Both are potentially useful for upscaling local measurements and quantifying large-scale carbon fluxes, especially if coupled to carbon-cycle models. These methods are very applicable in quantifying C fluxes in areas with no ground information, for example (for example in many tropical areas). However, their accuracy is often questioned. Another problem is that they are applicable to quantify carbon pools below ground (for example because of the number of processes involved and the spatial variation). In addition, validation by ground-based estimates is required to come up with reliable C-sink potentials.

Atmospheric *inversions* constrain the magnitude of terrestrial carbon sinks, but their ability to discern the responsible mechanisms or the exact location of the observed sink is limited.

There are different *modelling approaches* to quantify ecosystem dynamics and related C fluxes. Process-based models, for example, can explore the importance of the ecosystem's physiological responses to climate variability or increasing CO_2. But they often focus on specific segments without considering natural or human-induced disturbances in realistic detail. In contrast, models of land-use change focus on the effects of human land use and are insensitive to changes in ecosystem physiology. In general, the use of models to estimate changes in C fluxes may lead to less transparency because models differ and are often very complex and difficult to understand.

Source: Based on Houghton et al. (1999), Nabuurs et al. (1999) and Watson et al. (2000).

3.2   The Potential of C Sequestration and Biomass to Mitigate Climate Change

The different accounting systems, methods and indicators make an accurate quantification of the current and future potential for carbon sequestration in the biosphere difficult. In this section, we discuss various estimations of the C-sink potential and the potential to offset emissions through biomass energy, as given in the literature.

First, it is important to discuss the role of the terrestrial biosphere in the global carbon cycle, since this determines the natural potential for carbon uptake. The net uptake of the terrestrial biosphere shows significant diurnal, seasonal, year-to-year and even centennial dynamics. In the last two decades, the biosphere served as a sink (Table 8.1), taking up approximately a quarter of the carbon released from fossil-fuel combustion (Schimel et al., 1996). There is an indication that this uptake is specifically triggered by indirect human activities (especially the regrowth of young forests after deforestation during the first half of the 20th century, in combination with CO_2 and nitrogen fertilization (Dolman et al., 2001)) and that it can be maintained for a number of decades. There is, however, also an indication that this uptake may diminish around 2100 (Cramer and Field, 1999). This, in turn, has consequences for future commitment periods in which atmospheric concentrations of CO_2 should be stabilized.

*Table 8.1   Global CO_2 budget for the periods 1980–1989 and 1990–1999
(Gt C yr-1)*

Carbon Flux	1980–1989	1990–1999
1 Emissions from fossil-fuel combustion and cement production	5.4± 0.3	6.3± 0.4
2 Land-use emissions	1.7 (0.6– 2.5)	NA (1.6± 0.8)[1]
3 Storage in the atmosphere	3.3± 0.1	3.2± 0.1
4 Uptake by the ocean	1.9± 0.6	1.7± 0.5
5 Net uptake by terrestrial biosphere (=[3+4]–1)	–0.2± 0.7	–1.4± 0.7 (–0.7± 1.0)[1]
Total terrestrial uptake (= 5–2)	–1.9 (–3.8– 0.3)	NA (–2.3± 1.3)[1]

Note: 1. According to Prentice et al. (2001) insufficient data (NA) are available. As an indication the numbers of Watson et al. (2000) for the period 1989–1998 are presented in brackets.

Source: Prentice et al. (2001) and Watson et al. (2000).

Many studies show that, in contrast to the natural sink, the potential for a human-induced carbon sink may be large in the future. This would imply that a large part of the Kyoto targets could be fulfilled by using sinks. However, the sink potential depends on the time horizon, definitions and eligible categories of land-use options and indicators chosen.

The first issue is the time horizon, that is whether a short- or a long-term perspective is considered (Figure 8.1). For the short term (for example up to the first commitment period) the sink capacity under Article 3.3 (that is forestation activities) may be limited in most Annex B countries (some countries even have an account for a source of C). Areas for planting forest will rarely become available and it takes years before new forests store a significant amount of carbon, while even a small area of deforestation results in debits. In the long term it could be efficient to plant forests. Studies indicate that vast areas will become available for planting forests in Europe and North America up to 2050 because of agricultural and environmental policies. Forests planted in these areas would store significant amounts of carbon. In addition, the carbon storage in the soil compartment of a Kyoto forest could increase for decades.

A second issue in determining the potential domestic sink in Annex B countries is the type of activities that can be accounted for under Article 3.3 (ARD) and 3.4 (that is additional activities). The strict case of limiting

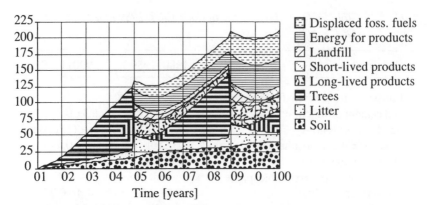

Notes: These are net changes in that, for example, the diagram shows savings in fossil-fuel emissions with respect to an alternate scenario that uses fossil fuels and alternative, more energy-intensive products to provide the same services (adapted from Marland and Schlamadinger, 1999; Kauppi et al., 2001).

Figure 8.1   Schematic representation of the cumulative carbon-stock changes and avoided emissions for a scenario involving afforestation and harvest

activities only to ARD significantly reduces the sink capacity (Table 8.2), because of the reasons mentioned above. Allowing activities under Article 3.4 as well increases the magnitude of the carbon sink. The sink potential depends on decisions about the categories of activities allowed (see Figure 8.2 and Box 8.4). If all activities are allowed, the potential in Annex B countries is estimated to be about 0.5 and 1 Gt C yr[-1] in 2010 and 2040, respectively (Watson et al., 2000). Analysing the individual activities shows that for the next 50 years, changes in forest management could result in an additional uptake of 0.5 Gt C yr[-1] (Watson et al., 2000); better management of agricultural soils could result in 0.4–0.6 Gt C.yr[-1] (Cole et al., 1996; Batjes, 1999; Lal and Bruce, 1999). Finally, allowing sinks under CDM would also considerably increase the potential of sinks (Table 8.2, Section 6). An issue related to the discussion about the carbon-sequestration activities under Articles 3.3 and 3.4 is the use of biomass as a source for energy or as a substitute for more energy-intensive materials. Such biomass use can also significantly contribute to a reduction of fossil-fuel emissions (Hall et al., 1991; Brown et al., 1996; Marland and Schlamadinger, 1997). The use of abandoned forest products for energy rather than disposal as waste could provide additional opportunities for displacing the use of fossil fuels (Apps et al., 1999). Recent studies indicate that in the next 30 to 40 years, the total potential for biomass production could increase up to 300–400 EJ yr[-1] (Faaij

BOX 8.4   EFISCEN

The European forest information scenario (EFISCEN) model is an area-based matrix model used to address, among other things, the impact of climate change and the potential carbon sequestration of European forests. With EFISCEN, insights can be gained in the development of forests in Europe (Nabuurs, 2001). The forests have been replanted in the course of the 19th and 20th centuries and are now mostly under 80 years of age. It is the history of these forests that determines their present and future capacity as a carbon sink. Figure B1 shows the effect of the vegetation rebound in terms of age-class development.

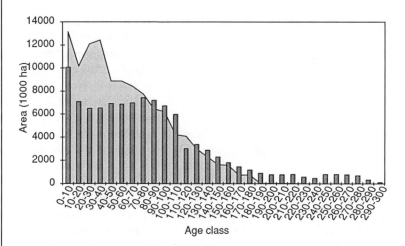

Figure B1   Age-class distribution of European forests in 1990 (filled area) and in 2090 (bars) as simulated with the EFISCEN model (Nabuurs, 2001)

EFISCEN demonstrates that even without climate change, there is a continuous build-up of the growing stock (thus, *sink*), assuming a certain annual increment and felling level. The average growing stock is shown to increase from 137 m^3 ha^{-1} in 1990 up to 245 m^3 ha^{-1} in 2050. Climate change in Europe could lead to an enhanced build-up of growing stock, up to 282 m^3 ha^{-1} in 2050 (Figure B2). It is within this autonomous vegetation rebound that the additionality of Kyoto measures must be decided upon.

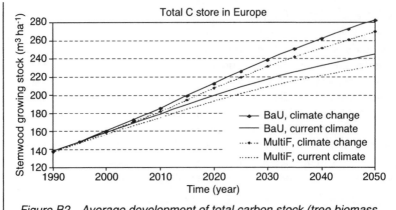

Figure B2   Average development of total carbon stock (tree biomass, soils and wood products) for European forests from 1990 to 2050 under two alternative scenarios of management and climate (BaU = Business as usual; MultiF = Multi functional management) (Kramer and Mohren, 2001)

and Agterberg, 2000; UNDP, 2000; Watson et al., 2000), which is close to the world's current energy production of approximately 400 EJ (resulting in emissions of about 5.4 Gt C yr^{-1}). The use of biomass to substitute for more energy-intensive materials may be limited with current policies (Karjalainen, 1996; Marland and Schlamadinger, 1997). Recent studies, however, show that the long-term effect would be large if new, effective policies to stimulate the use of wood as a basic material would be initiated (Gielen et al., 2001; Hekkert, 2000). In addition to the potential contribution of biomass, the advantages of using biomass are its relatively even geographical distribution and its relatively low price (see next section). In addition, it does not have some of the disadvantages of carbon sequestration (for example permanence is not a question because biomass substitutes for carbon emissions from the use of fossil fuels), although some other disadvantages remain (for example the competition for land).

A last topic that determines the net land use related carbon potential is whether or not such projects are acceptable in non-Annex B countries through CDM (see Section 6). The issue of accounting for CDM projects is controversial because, on one hand, the potential sink is huge (Table 8.2, Box 8.5), but on the other hand, there are many disadvantages (see also Section 5). Potentially 0.5–1.4 Gt C yr^{-1} could be sequestered in non-Annex B countries in the next 50 years (Onigkeit et al., 2000; Watson et al., 2000).

Table 8.2   Range of estimated potential of most important sink categories for the first commitment period and comparison to estimated differences between baselines and assigned amounts for OECD countries

Category	Min potential (MtC.yr⁻¹)¹	Max potential (MtC. yr⁻¹)	Remarks
Art 3.3	5	140	High end due to application of Art 3.7 by some countries
Art 3.4	50	500	Based on IPCC assumptions about percentage of lands covered for various land-use activities; extending it to large parts of the land will drastically increase high amount
Sinks in CDM	50	130	High end based on IPCC assumptions about percentage of lands in developing countries where land-use activities could be undertaken (afforestation, reforestation and land-use practices) and a conservative estimate that only 10 per cent of that potential can be tapped by project-based activities during the first commitment period; bringing more land in or increasing the project area can significantly increase the high amounts; it does not include avoided deforestation (good for a maximum of 1600 MtC/yr) given that this is strongly opposed by many countries
OECD	750	1200	This is the estimated difference 'demand' between baseline emissions in the commitment period and assigned amounts without any other action to close that gap (Vrolijk et al., 2000)

Note: 1. According to IPCC Special Report on LULUCF (Watson et al., 2000) and Noble and Scholes (2000).

Source: Metz et al. (2001), Watson et al. (2000), Vrolijk et al. (2000), Noble and Scholes (2000).

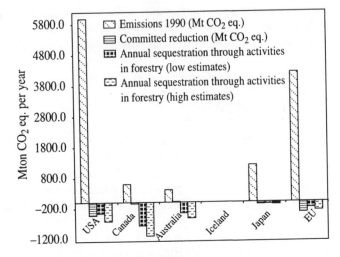

Notes: Estimates include all applicable forestry activities and assume adoption on average of about 10 per cent of available area. Estimates are derived from the 'access to country-specific data' (ACSD) tool (see also Box 8.5). The values are presented for illustration only and vary among assumptions used and chosen set of measures. Values for the four bars for Iceland are 2.6, –0.1, –0.2 and –0.3, respectively (adapted from Nabuurs et al., 2000).

Figure 8.2   *Potential carbon sink in forests for selected countries and forest-management activities under Article 3.4 (Mt CO$_2$ equivalents), relative to commitments and assigned amounts under the Kyoto Protocol*

BOX 8.5   IMAGE 2

IMAGE 2 (IMAGE stands for 'integrated model for assessing the global environment') is an integrated model for assessing global change. It explicitly simulates, among others, the dynamics of land use and land cover. Changes in land use are driven by changes in climate and in demands for food, fodder, wood, fuel and carbon sequestration. These demands are satisfied by evaluating current land-use patterns and the potential production of different uses (for example agricultural, forest). Based on the potential production and demand the model determines where land-use categories emerge or abandon (for details, see Alcamo et al., 1998).

The terrestrial land cover consists of 18 classes. Among these are classes for natural vegetation, agricultural land uses and 'Kyoto forest' (that is forest plantations that are planted under the Kyoto Protocol). Kyoto forests may appear though afforestation of new areas or reforestation of areas that are already covered by forest (after the harvest). Since the model also deals with deforestation and related greenhouse gas emissions, the model is suitable for evaluating the sink potential under Article 3.3 of the Kyoto Protocol and the consequences of including ARD activities in CDM.

Ten different types of Kyoto forests have been defined in IMAGE to represent the variation in plantations. The types vary in regard to optimal climatic conditions, growing characteristics, and so on. The potentially best-growing Kyoto forest will be planted in an area, using a set of environmental constraints and the carbon dynamics as the driving force. The computation of the carbon dynamics of Kyoto forests is similar to the model of the carbon cycle used for natural types of vegetation (Klein Goldewijk et al., 1994). In addition, the so-called 'surplus primary productivity' (SPP) is computed to determine whether planting a Kyoto forest is worthwhile in the context of reducing atmospheric CO_2 (see Section 3 below and Onigkeit et al., 2000).

The IMAGE 2 model has been used to determine the sink potential for an A1 and B1 world (Nakicenovic, 2000). Basic considerations in the evaluation include different options for land that can be forested and the use of the wood from the Kyoto forests. The latter is important because an efficient use of wood could result in additional reductions of the atmospheric CO_2 concentration. The comparison of different options depicts the importance of including ARD activities under CDM and possible conflicts with other land-use functions like food security and biodiversity.

The application of IMAGE 2 demonstrates that planting Kyoto forests can be an effective policy option from the point of CO_2 reduction, even without jeopardizing food security. However, much of this – as well as possible side effects like loss of biodiversity – depends on the definitions used (for example which land can be used) and whether ARD activities are part of CDM. Potentially, about 1 to 8 million km^2 of Kyoto forests can be planted up to the year 2100 (globally). The lower edge of the range reflects the potential when Kyoto forests are allowed only on abandoned agricultural land, with plantations planted in both Annex B and non-Annex B countries. Natural systems are seldom threatened under this assumption, but they are also less effective in reducing atmospheric CO_2. Thus reductions in fossil fuel emissions possibly combined with additional activities under Article 3.4 are needed. The contribution of Kyoto forests to an effective reduction in CO_2 could be potentially considerably larger if the forests could also be planted in natural areas. Mainly savanna and natural grassland area will be used. Forests are seldom converted, since they have a negative SPP. The largest potential has been computed for the case that plantations could be planted in different natural areas also in non-Annex B countries. Thus, assuming that many types of forestation activities are accepted under CDM, the effect on the atmospheric CO_2 concentration of foresting such vast areas would be large. In a B1 world, for example, Kyoto forests could account for half of the commitments to reach stabilization at 450 ppm CO_2. However, such vast Kyoto plantations could threaten biodiversity and social structures, and the extra management activities needed to enable the establishment of the plantations (such as fire suppression) would incur additional costs (which are not included in our analysis).

The afforestation/reforestation activities would be even more effective if a cascade of different uses were allowed (i.e., first sequestration, subsequently used to meet the demand for either timber of biofuels). Up to the year 2100, the extra gain would vary between 4 per cent and 15 per cent of the needed reduction, depending on the baseline (A1 or B1) and definitions (only abandoned agricultural land or also natural systems).

3.3   Side Effects of Carbon Conservation, Sequestration and Biomass Activities

Activities to stimulate carbon conservation, sequestration or fuel substitution provide significant socio-economic and environmental side effects (Makundi, 1997; Brown, 1998; Watson et al., 2000). These side effects do not directly determine the sink potential of an activity (that is, the net effect on atmospheric CO_2 concentrations), but they could contribute to the discussion about the usefulness of an activity.

Side effects can be positive or negative. Often, the net result depends on the design and implementation of the project (WBGU, 1998; Watson et al., 2000). Further, the result depends also on the type of activity. Carbon conservation, for example, often improves biodiversity, whereas afforestation activities may negatively affect the biodiversity in a region because of the relative low number of species they support and the introduction of exotic species. Only if the plantations are established in grassland areas or degraded land can the net effect on biodiversity be limited. Similar, carbon conservation often results in less deforestation. The effect of ARD projects on the deforestation in a region depends on the original land cover, whether or not the project starts with the harvest of virgin forests (WBGU, 1998).

In general, the positive effects of many carbon-sequestration and biomass activities are the potential to improve soil properties, to protect watersheds and to reduce the pressure on remaining natural forests. Further, land-use activities may be accompanied by the introduction of new land-use activities and technologies, promoting sustainable agricultural practices. A particularly positive side effect associated with carbon-substitution projects is the possibility for local communities to get decentralized, more stable, energy-supply systems.

The negative environmental and socio-economic impacts associated with land-use activities are often related to their possible impact on the local community. For example, afforestation activities may compete for land for which a community has other priorities (such as agricultural production, Lugo, 1997). Further, concern is often expressed because of the non-permanence of afforestation activities.

To summarize, activities aimed at carbon conservation, carbon sequestration or fuel substitution are neither inherently good nor bad in terms of possible side effects. The net result strongly depends on their design and implementation, the type of activity and the involvement of the local community.

4   ECONOMIC ASPECTS

The main economic issues on offsetting emissions through carbon sinks or energy from biomass concern the potential change in emissions and the costs of the various options, including ancillary costs and benefits. Since the potential and the costs differ for various regions in the world, an important issue is whether the low-cost options that are available in developing countries and Eastern Europe can be used through the clean development mechanism (CDM) and joint implementation (JI) (see next section).

The economic costs of terrestrial sinks and energy from biomass are calculated on the basis of the required investment costs and the annual operation and maintenance costs of afforestation, reforestation, reduced deforestation or farm management, including the costs of alternative uses of the land. In comparing alternative options, it is necessary to estimate costs and revenues over the period of the projects involved and to discount to present values.[1] For agroforestry, the growth characteristics of the forest and the rotation period are considered in addition to the life time of a project, and assumptions are needed about the use of wood products, their life time and the waste-management systems at the end of their life cycle.

On the basis of various case studies (see Box 8.6 for an example for Mexico), cost functions have been constructed, indicating the potential and cost of various options for reducing or offsetting greenhouse gas emissions by means of carbon sinks or biomass energy.

BOX 8.6   MEXICO – AGROFORESTRY

A case study on forestry and agroforestry in the Central Highlands of Chiapas in Mexico is presented in De Jong (2000). He analyses the costs of reducing GHG emissions by forestation and management of fallow land, agricultural land and pasture in Mexico. The study considers the private costs of these management options and the opportunity costs of agricultural income foregone. It pays extensive attention to the risk of carbon leakage, the possibility that additional deforestation will take place at another location, outside the project boundaries.

The study shows that reforestation can result in large reductions of CO_2 emissions at relatively low costs (US$5–20 per ton of carbon). Within this range, measures involving forestry and agroforestry in the relevant study area could mitigate from 1 Mt C to 42 Mt C, with a maximum economic supply of carbon sequestration of about 55 Mt C at US$40 per ton C. If indeed the forestation project is additional and no carbon leakage takes place, the project would be useful for CDM. However, it is not clear whether reforestation would occur anyway or not at all if no climate policies were implemented. It is therefore difficult – if not impossible – to make a scientific judgment about the additionality of the project.

4.1 Costs of Biomass

Biomass contributes significantly to the world's energy supply, probably accounting for about 45 EJ a year (9–13 per cent of the world's energy supply). Estimates for the contribution of biomass to future energy demands shows that biomass may play an important role (up to 300–400 EJ yr^{-1}, see previous section), thus contributing to offsetting emissions from fossil fuels. It is estimated that with agriculture modernized up to reasonable standards in various regions, and given the need to preserve and improve the world's natural areas, 700–1400 million hectares may be available for biomass production well into the 21st century (UNDP, 2000). The economic costs of biomass energy systems differ from region to region, depending on the price of land, labour and capital, and the local efficiency of technologies and management. Prices range from US$1.5–2.0 per GJ in Brazil, to US$4 per GJ in some parts of Europe. With biomass prices of about US$2 per GJ and state-of-the-art combustion technology at a scale of 40–60 megawatts of electricity, biomass could result in electricity production costs of about US$0.05–0.06 per kilowatt-hour (UNDP, 2000). Whether large-scale biomass energy systems will become competitive depends on cost reductions through technological progress and the future level of carbon prices. Carbon prices in the range of US$20 to US$60 and above per ton of CO_2 equivalent would definitely provide strong incentives to further expand energy supplies from biomass in the USA, Europe and developing countries.

4.2 Carbon Sequestration and Biomass Energy in the USA

McCarl and Schneider (2000) use the Agricultural Sector Model to analyse the potential carbon emission sequestration in the agricultural sector in the USA at varying carbon price levels. They estimate a potential for the US of about 350 million metric tons (Mt) of carbon equivalent at a price of US$500 per ton C equivalent. However, at lower prices (in the range of US$50 or 100 per ton C equivalent), the potential remains around 125 Mt of carbon equivalent or less. Figure 8.3 shows the total carbon emissions avoided by the use of different land-use options. The figure also shows a relatively low potential for sequestration at carbon prices below US$60 per ton of carbon equivalents but a rapid increase for biomass for power plants. Although soil carbon sequestration can only make a relatively small contribution, it is an interesting possibility at low carbon prices.

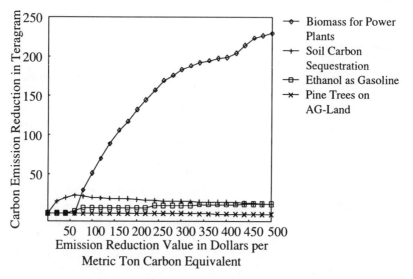

Source: McCarl and Schneider (2000).

Figure 8.3   Potential and costs of reducing carbon emissions in dollars per metric ton of carbon equivalent

4.3   Carbon Sequestration in Soils

Batjes (1999) indicates the global potential for sequestration through improved soil management from 580 to 800 MMT C per year over the next 25 years. There are only a few estimates regarding the costs of carbon sequestration in soils. The study of McCarl and Schneider (2000) reports a potential for the US of about 30 Mt of carbon equivalent for a carbon price of US$40 per metric ton of carbon equivalent and above (Figure 8.3). Further studies should reveal how much could be sequestered at what cost in other regions of the world.

4.4 Emission Offsets from Forestry Compared to Emission-reduction Options

How do emission offsets from forestry compare to other options for reducing emissions?

Although the different studies have been based on different assumptions, it is still interesting to compare the cost estimates reported for various types of land-use activities. Figure 8.4 shows that forestry in developing countries is a low-cost opportunity for large-scale carbon sequestration provided that transaction costs are kept within reasonable limits and that no excessive carbon leakage occurs (Kauppi et al., 2001). In the short term, fuel-switch also provides good opportunities at low costs. Figure 8.4 also shows that renewable energy is relatively expensive, and that forestry in OECD countries offers limited potential for offsetting CO_2 emissions at a low price (maximum amount that can be offset by forestry in OECD countries is 250 Mt CO_2 at a price of US\$50 per ton of CO_2). Thus, carbon sequestration and

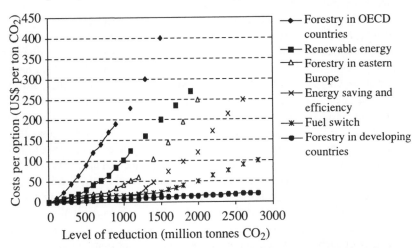

Notes: The curves show cost differences among world regions between comparable options. However, costs per option are also reported to vary considerably, and at comparable total levels of reduction. This is mainly because studies on costs have not been carried out in the same way. In some options, a net monetary profit may also be made (that is, costs may also be negative).

Source: Kauppi et al. (2001), based on Brown et al. (1996), Hol et al. (1999).

Figure 8.4   Indicative curves of costs (in US\$ per ton CO_2) of emission reduction or carbon sequestration according to level of total reduction

biomass could provide opportunities to reduce the net accumulation of carbon in the atmosphere at relatively low costs (in the range of US$0–50 per tonne of CO_2), but only for a relatively modest quantity.

5   SINKS AND BIOMASS ENERGY AS A PART OF JOINT IMPLEMENTATION (JI) AND THE CLEAN DEVELOPMENT MECHANISM (CDM)

The Kyoto Protocol offers opportunities for Annex B countries to achieve parts of their emission reductions through land-use measures in other Annex B countries (as part of joint implementation) or non-Annex B countries (as part of the clean development mechanism). Given regional cost differences, it is interesting to discuss the role of JI and CDM in using the options of carbon sinks and biomass energy. Because of differences in the cost of reducing emissions in various countries in the world, cost-effectiveness can be enhanced by means of these mechanisms. However, serious problems may occur with their practical implementation. CDM accounting projects, especially, have raised specific concerns. First, they are surrounded by large uncertainties that exist in basic data, the permanence of the carbon storage and the problem of additionality (that is crediting reductions that would have occurred anyway) and leakage (that is crediting sink activities, the gain of which becomes counterbalanced by carbon losses outside the project area). Second, the effectiveness of land-use projects under CDM is difficult to monitor and verify. Various problems remain for the actual implementation of such projects. Clear methods should be developed to establish the net additional reductions of the emissions of various projects. This holds both for projects focused on sustainable energy (like solar, wind, hydro and biomass) and for projects for carbon sequestration through afforestation, reforestation or reduced deforestation. Special attention will be required for the various methods used in establishing the baseline emissions and for tackling the possibility of carbon leakage, that is the option that additional greenhouse gas emissions will occur elsewhere as an undesired side effect of a project.

At present, more and more developing countries are showing an interest in participating in CDM projects because of the opportunities for additional financing of projects that foster economic growth and contribute to sustainable development. In the longer term, developing countries might have to fulfil their own targets for reducing greenhouse gas emissions, and there is a risk that under CDM, carbon sinks and biomass systems have been picked as the 'low-hanging fruits' by industrialized countries, with only very expensive options remaining for developing countries to fulfil their obligations in the second commitment period. In assessing the costs of projects under CDM, it is essential to include all cost aspects, including the cost of monitoring and the transaction costs of establishing the contracts. For example, for carbon sequestration in agroforestry a large number of farmers will have to be contracted and the transaction costs could become prohibitive. Monitoring of emission reductions is essential but expensive, particularly if the monitoring of all sinks in the relevant country becomes necessary to avoid carbon leakage.

6   POLICY RELEVANCE AND VIEWS OF PARTIES

In the previous sections, it was demonstrated that carbon sequestration and biomass activities offer the potential to many Annex B countries to meet their Kyoto commitments at relatively low cost. This potential is enhanced and prices are lower when JI and CDM projects are considered. However, there is also a significant amount of uncertainty and risk surrounding carbon-sequestration and biomass activities (for example because of the mentioned permanence, the difficulty of measuring the sink, and so on), which has resulted in different views among parties to the UNFCCC (ENB, 2000). The perception of some countries, such as the USA and Canada, was straightforward: their view was that existing forests and agricultural land contribute to carbon sequestration and that these should be taken into account in meeting their Kyoto targets, regardless of whether the uptake is due to natural causes or human activity. These countries proposed a comprehensive, broad-based approach, among other things, because this would best reflect the net effect on the atmosphere. They supported the inclusion of all possible activities under Article 3.4 and an inclusion of sinks in the CDM.

Other countries, like the EU member countries, also recognized the large potential (Metz et al., 2001), but they were more restrained in including sinks and biomass options in meeting commitments for several reasons. First, as mentioned in Section 2, the potential is significantly greater for some countries, which could lead to a very uneven distribution of benefits. Second, the permanence of the activities was a major concern. The additional carbon sequestration in the biosphere may diminish if, for example, the frequency and/or extent of disturbances, such as forest fires, were to change under future land-use activities and variations in climate (Cramer and Field, 1999). Furthermore, uncertainties due to leakage and additionality have been a major concern. While in JI projects both issues would automatically be taken care of (through national inventories), they might be difficult to detect and avoid under CDM. Third, the verification of sinks under CDM was questioned.

Finally, the countries were specifically against the inclusion of ecosystem conservation in CDM. Because of the huge potential (about 1.6 Gt C yr^{-1}) of including ecosystem conservation, it could 'flood' the market, even if only a portion would be brought under CDM.

The position of many developing countries (organized in the so-called 'G77 and China' group) is divided. On the one hand, many take the view that Annex B countries should take the responsibility of reducing their domestic greenhouse gas emissions (without picking 'low-hanging fruits', as mentioned in the previous section). On the other hand, some countries, particularly those from Latin America, are open to including direct, human activities in CDM. These countries see these activities as a potential way to gain financially from CDM projects.

Various proposals have been made to bridge the polarization of views. For example, the EU proposed restricting categories under Article 3.4 during the first commitment period, but might accept changes at a later stage, after the implications have become clear (Schlamadinger and Marland, 2000; Metz et al., 2001). Another proposal was made to restrict activities under Article 3.4 (so-called 'discounting'), whereas there would be no cap on ARD projects. Other suggestions were made to leave forestry-related activities out of CDM projects, but bring in similar projects via different routes (Metz et al., 2001). The G77 and China group, for example, proposed accepting forest conservation and the reforestation of degraded land as an activity for adapting to climate change (for which funds are available, see ENB, 2000). Another route would be to award a more prominent role to biomass as a substitute for fossil fuel and energy-intense materials. As mentioned above, biomass projects do not have the specific problems of carbon sequestration (for example permanence and verification).

Major progress has been made at COP 6 bis in Bonn in dealing with these different concerns. This includes agreements on exactly how countries (with the exception of the US) may include measures related to land-use:

- An agreement on the definition of 'forests'; and the activities 'afforestation', 'reforestation' and 'deforestation'. These activities shall be defined on the basis of a change in land use.
- Sinks as a result of ARD activities can be fully used.
- Sinks due to forest management can be fully used, but only up to a maximum of 8.2 Mt C yr^{-1} and only up to the level of debits due to ARD (for some countries the net effect of ARD is currently negative, that is an extra source of carbon).
- Additional management (Article 3.4) can be fully used to achieve the commitment; however, this is only up to an absolute amount determined

for each country (for example 1.24 Mt C yr^{-1} for Germany). For some countries, these amounts are relatively large, whereas for others, they are rather small.

- Each country may choose to apply any or all of the activities during the first commitment period, as agreed in Article 3.4.
- Agricultural activities under Article 3.4 (for example cropland management) should be implemented by applying the net–net accounting approach (that is net emissions or removals during the commitment period minus removals in the base year, see Section 3, above).
- The ongoing aging effect in many forests (especially in the northern hemisphere) and the CO_2 and N fertilization effect should be excluded from accounting.
- JI projects can also be used to achieve the commitment, but with the same restrictions as above.
- Only projects falling under afforestation or reforestation can be used in developing countries (as part of CDM projects). Thus, ecosystem conservation (for example avoiding deforestation) is excluded. The CDM projects can only be used up to a maximum of 1 per cent of the 1990 emissions of a particular Annex-B country.

Thus, an agreement was established among the parties (with the exception of the USA) about the definitions, type and extent of activities allowed under Articles 3.3 and 3.4 and how sinks and biomass projects can be applied under JI and CDM. Despite these achievements, parties still have to work out details about the preferred approach. Furthermore, some issues remain open. This includes the development of definitions for afforestation and deforestation under CDM in the first commitment period, taking into account issues of permanence, additionality and age, as well as socio-economic and environmental impact. The Subsidiary Body for Scientific and Technological Advice (SuBSTA) has been asked to provide additional advice.

7   DISCUSSION AND CONCLUSIONS

Carbon sequestration and biomass offer the opportunity to slow down the increase in atmospheric CO_2 concentration considerably. It is possible to store up to about 1 to 2 Gt C yr^{-1} in vegetation (especially trees) and soils of the terrestrial biosphere and related products. The actual sink depends mainly on the definitions, the eligible categories of land-use management chosen and whether or not sink projects are allowed under CDM. With respect to biomass, studies have shown that in the coming decades the potential for biomass production may increase up to a level equivalent to the current global energy demand. Whether large-scale biomass systems will become competitive depends on the implementation of policies, cost reduction through, for example, technological progress, and the future level of carbon prices. An efficient combination of sequestration and biomass involves first sequestering carbon in wood products and, at the end of their lifecycle, using the biomass for energy recovery.

An important factor that determines the net potential of carbon sinks and biomass as an option for climate change mitigation is whether or not JI will become successful and sinks will be allowed under CDM. In addition to the need for political decisions about these issues, success depends on the willingness of industrialized countries to finance projects in Eastern Europe or in developing countries. The costs of such projects are often limited. Furthermore, technology transfer and the consideration of local conditions can increase the acceptance of carbon sequestration and biomass projects under JI and CDM projects.

If applied properly, large-scale carbon sequestration and biomass production have various ancillary socio-economic and environmental benefits. Improved land management (as an Article 3.4 activity), for example, can contribute both to carbon sequestration and improvement of agricultural soils, leading to higher yields. Furthermore, as mentioned before, the chances of success for sink and biomass projects increases if investments are made in local communities. The environment may benefit from sequestration and biomass projects through increased interest in avoiding soil erosion and improving land-cover protection (with positive effects on biodiversity and watersheds). Because of these positive side effects, sink and biomass projects may contribute to sustainable development.

Despite the large potential of carbon sinks and biomass as options to mitigate climate change and the side benefits, there are still difficulties that should be considered before actual implementation becomes feasible. First, it could be argued that sinks and biomass projects have not only positive side effects, but that there is also the risk of negative environmental and social side effects. For example, large-scale afforestation could lead to negative effects on biodiversity in a region because of a decreased richness of species and the introduction of exotic species. Careful organization of a sink or biomass project could address these negative side effects (for example, by limiting large-scale monoculture plantations). Second, reliable monitoring and verification systems are needed. Although not yet completely developed, these systems might prove to be less difficult to establish because of the progress that has been made in the development of monitoring mechanisms. Finally, we have stressed in this chapter that sink and biomass projects are a concern to many because of relatively large uncertainties, especially if they are implemented under CDM. Issues like the permanence, leakage and additionality have been a concern. The system of monitoring, verification and accounting should be managed in such a way as to make beneficiaries fully aware of and responsible for lower carbon results than projected. The recent implementation, as accepted by the parties to the UNFCCC (with the exception of the USA), addresses parts of these uncertainties, by including measures for discounting carbon sinks in CDM, for example. Nevertheless, some issues remain unsolved (for example how to deal with leakage).

In summary, the implementation of activities to either sequester carbon or replace fossil-fuel use requires careful analysis of economic and technological conditions along with the various local options available for carbon sequestration and biomass. Some hurdles have still to be overcome before measures can be implemented. If, however, these problems are resolved and solutions are successfully implemented, carbon sequestration and biomass production could play an important role in reducing net greenhouse-gas emissions. In addition, there are a number of ancillary benefits for local communities and the environment.

NOTE

1. The net present cost of a project can be calculated as $NPC = I_0 + \sum [e^{-rt} * (C_1 - B_1 + OC_1)]$ where NPC = net present cost, I_0 = investment in base year, r = discount rate, C = annual operations and maintenance costs, B = annual benefits, OC = annual opportunity costs, and t = time $(O, ..., T)$. The costs per tonne can then be calculated as the NPC divided by the total net quantity of carbon stored in the sink over the lifetime of the project.

REFERENCES

Alcamo, J., R. Leemans and G.J.J. Kreileman (1998), *Global Change Scenarios of the 21st Century. Results from the IMAGE 2.1 Model*, London: Pergamon Press/Elsevier Science.

Apps, M.J., W.A. Kurz, S.J. Beukema and J.S. Bhatti (1999), 'Carbon Budget of the Canadian Forest Product Sector', *Environmental Science and Policy*, 2, 25–42.

Baldocchi, D., R. Valentini, S. Running, W. Oechel and R. Dahlman (2000), 'Strategies for Measuring and Modelling Carbon Dioxide and Water Vapour Fluxes over Terrestrial Ecosystems', *Global Change Biology*, 2, 159–68.

Batjes, N. (1999), *Management Options for Reducing CO$_2$-Concentrations in the Atmosphere by Increasing Carbon Sequestration in the Soil*, Bilthoven, The Netherlands: Dutch National Research Programme on Global Air Pollution and Climate Change, Report No. 410200031.

Brown, S. (1998), *Climate, Biodiversity and Forests: Issues and Opportunities Emerging from the Kyoto Protocol*, Washington DC, USA: World Resource Institute.

Brown, S., J. Sathaye, M. Canell, P. Kauppi, P. Burschel, A. Grainger, J. Heuveldop, R. Leemans, P. Moura Costa, M. Pinard et al. (1996), 'Management of Forests for Mitigation of Greenhouse Gas Emissions', in R.T. Watson, M.C. Zinyower, and R.H. Moss (eds), *Climate Change 1995. Impacts, Adaptations and Mitigation of Climate Change: Scientific-Technical Analysis*, Cambridge: Cambridge University Press, pp. 773–97.

Cole, V., C. Cerri, K. Minami, A. Mosier, N. Rosenberg, D. Sauerbeck, J. Dumanski, J. Duxbury, J. Freney, R. Gupta et al. (1996), 'Agricultural Options for Mitigation of Greenhouse Gas Emissions', in R.T. Watson, M.C. Zinyowera and R.H. Moss (eds), *Climate Change 1995. Impacts, Adaptations and Mitigation of Climate Change: Scientific-Technical Analysis*, Cambridge: Cambridge University Press, pp. 745–71.

Cramer, W. and C.B. Field (1999), 'Comparing Global Models of Terrestrial Net Primary Productivity (NPP): Introduction', *Global Change Biology*, **5**, 3–5.

De Jong, H.J. (2000), *Forestry for Mitigation of the Greenhouse Effect: An Ecological and Economic Assessment of the Potential of Land Use to Mitigate CO$_2$ Emissions in the Highlands of Chiapas, Mexico*, PhD thesis, Wageningen, The Netherlands: Wageningen University.

Dolman, H., G.J. Nabuurs, P. Kuikman, B. Krijt, S. Brinkman, L. Vleeshouwer and J. Verhagen (2001), *Terrestrial Carbon Sinks and the Kyoto Protocol: The Scientific Issues*, Bilthoven, The Netherlands: Dutch National Research Programme on Global Air Pollution and Climate Change, Research Report.

ENB (2000), Summary of the 6th Conference of Parties, *Earth Negotiations Bulletin*. Available on-line at http://www.iisd.ca/vol12/enb12163e.html.

Faaij, A. and A. Agterberg (2000), *Long-term Perspectives for Production of Fuels from Biomass; An Integrated Assessment and R&D Priorities*, Department of Science, Technology and Society, Utrecht University, The Netherlands, paper prepared for the 4th Biomass Conference, Oakland, 1999, Utrecht,

Fan, S., M. Gloor, J. Mahlman, S. Pacala, J. Sarmiento, T. Takahashi and P. Tans (1998), 'A Large Terrestrial Carbon Sink in North-America Implied by Atmospheric and Oceanic Carbon Dioxide Data and Models', *Science*, **282**, 441–46.

Gielen, D.J., M.A.P.C. De Feber, A.J.M. Bos and T. Gerlagh (2001), 'Biomass for Energy or Materials? A Western European System Engineering Perspective', *Energy Policy*, **29**, 291–302.

Goudriaan, J. (1995), 'Global Carbon Cycle and Carbon Sequestration', in M. Beran (ed.), *Prospects for Carbon Sequestration in the Biosphere*, NATO ASI series, pp. 3–18.

Hall, D.O., H.E. Mynick and R.H. Williams (1991), 'Cooling the Greenhouse with Bioenergy', *Nature*, **353**, 11–12.

Hekkert, M.P. (2000), *Improving Material Management to Reduce Greenhouse Gas Emissions*, PhD, Utrecht: Utrecht University, pp. 217.

Hol, P., R. Sikkema, E. Blom, P. Barendsen and W. Veening (1999), *Private Investments in Sustainable Forest Management*, Wageningen, The Netherlands: Form Ecology Consultants and Netherlands Committee for IUCN, Final Report.

Houghton, R.A., J.L. Hackler and K.T. Lawrence (1999), 'The US Carbon Budget: Contributions from Land-use Change', *Science*, **285**(5427), pp. 574–78.

Karjalainen, T. (1996), 'Dynamics and Potentials of Carbon Sequestration in Managed Stands and Wood Products in Finland under Changing Climatic Conditions', *Forest Ecology and Management*, **2**, 165–75.

Karjalainen, T., A. Pussinen, J. Liski and G.J. Nabuurs (2001), 'Impacts of Forest Management and Climate on the Forest Sector Carbon Budget in the Nordic Countries', in M. Apps et al. (eds), *The Role of Real Forests in the Global Carbon cycle-Proceedings*, Edmonton.

Kauppi, P., R.J. Sedjo, M. Apps, C. Cerri, T. Fujimori, H. Janzen, O. Krankina, W. Makundi, G. Marland, O. Masera et al. (2001), *Technical and Economic Potential of Options to Enhance, Maintain and Manage Biological Carbon Reservoirs and Geo-engineering*, Cambridge University Press, UK: IPCC Third Assessment Report on Climate Change, Chapter 4 WG III.

Klein Goldewijk, K., J.G. Van Minnen, G.J.J. Kreileman, M. Vloedbeld and R. Leemans (1994), 'Simulation of the Carbon Flux Between the Terrestrial Environment and the Atmosphere', *Water, Air and Soil Pollution*, **76**, 199–30.

Kramer, K. & G.M.J. Mohren (2001), *Long-term regional effects of climate change on European forests: impact assessment and implications for carbon budgets*, Final report of LTEEF II, ALTERRA Report 194, ALTERRA, Wageningen.

Lal, R. and J.P. Bruce (1999), 'The Potential of World Cropland Soils to Sequester C and Mitigate the Greenhouse Effect', *Environmental Science and Policy*, **2**, 177–85.

Lugo, A. (1997), 'An Apparent Paradox of Re-establishing Species Richness on Degraded Lands With Tree Monocultures', *Forest Ecology and Management*, **99**, 9–19.

Makundi, W.P. (1997), 'Global Climate Change Mitigation and Sustainable Forest Management: The Challenge of Monitoring and Verification', *Mitigation and Adaptation Strategies for Global Change*, **2**, 133–55.

Marland, G. and B. Schlamadinger (1997), 'Forests for Carbon Sequestration or Fossil Fuel Substitution? A Sensitivity Analysis', *Biomass and Bioenergy*, **13**, 389–97.

Marland, G. and B. Schlamadinger (1999), 'The Kyoto Protocol Could Make a Difference for the Optimal Forest Based CO_2 Mitigation Strategy: Some Results for GORCAM', *Environmental Science & Policy*, **2**, 111–24.

Matthews, R., G.J. Nabuurs, V. Alexeyev, R. Birdsey, A. Fischlin, J.P. MacLaren, G. Marland and D. Price (1996), 'Evaluating the Role of Forest Management and Forest Products in the Carbon Cycle', in M. Apps and D. Price (eds), *Forest Ecosystems, Forest Management and the Global Carbon Cycle*, Heidelberg: Springer, pp. 293–301.

McCarl, B. and I. Schneider (2000), *Economic Potential of Biomass Based Fuels from Forests and Agricultural Sources*, Conference on Sustainable energy: New challenges for agriculture and implications for land use, Wageningen University, 18–20 May, 2000, Wageningen, The Netherlands.

Metz, B., M. Berk, M. Kok, J.G. Van Minnen, A. De Moor and A. Faber (2001), 'How Can the European Union Contribute to a CoP-6 Agreement? An Overview for Policy Makers', *International Environmental Agreement: Politics, Law and Economics*, **1**, 167–85.

Mitchell, J.F.B., D.J. Karoly, G. Hegerl, F.W. Zwiers and J. Marengo (2001), *Detection of Climate Change and Attribution of Causes*, IPCC Third Assessment Report on Climate Change, Chapter 12 WGI, Cambridge University Press.

Nabuurs, G.J. (2001), *European Forests in the 21st Century: Impacts of Nature-oriented Forest Management Assessed with a Large-scale Scenario Model*, PhD thesis, University of Joensuu, Finland.

Nabuurs, G.J., R. Paivinen, R. Sikkema and G.M.J. Mohren (1998), 'The Role of European Forests in the Global Carbon Cycle – A Review', *Biomass and Bioenergy*, **13**, 345–58.

Nabuurs, G.J., A.J. Dolman, E. Verkaik, A.P. Whitmore, W.P. Daamen, O. Oenema, P. Kabat and G.M.J. Mohren (1999), *Resolving Issues on Terrestrial Biospheric Sinks in the Kyoto Protocol*, Bilthoven, The Netherlands: Dutch National Research Programme on Global Air Pollution and Climate Change, Report No. 410 200 030.

Nabuurs, G.J., A.J. Dolman, E. Verkaik, P.J. Kuikman, C.A. van Diepen,

A.P. Whitmore, W.P. Daamen, O. Oenema, P. Kabat and G.M.J. Mohren (2000), 'Article 3.3 and 3.4 of the Kyoto Protocol: Consequences for Industrialised Countries' Commitment, the Monitoring Needs, and Possible Side Effects', *Environmental Science and Policy*, **3**, 123–34.

Nakicenovic, N. et al. (eds) (2000), *IPCC Special Report on Emissions Scenarios*, Cambridge University Press.

Noble, I.R. and R.J. Scholes (2000), 'Sinks and the Kyoto Protocol', *Climate Policy*, **1**, 5–25.

Onigkeit, J., M. Sonntag and J. Alcamo (2000), *Carbon Plantations in the IMAGE Model – Model Description and Scenarios*, Kassel, Germany: Center for Environmental Systems Research, University of Kassel, COOL project report.

Prentice, I.C., M.J.R. Farquah, M.J.R. Farham, M.L. Goulden, M. Heinmann, V.J. Jaramillo, H.S. Kherghi, C. Le Quere, R.J. Scholes and D.W.R. Wallance (2001), *The Carbon Cycle and Atmospheric Carbon Dioxide*, Cambridge University Press: IPCC Third Assessment Report on Climate Change, Chapter 3 WGI.

Sarmiento, J.L. and S.C. Wofsy (1999), *A U.S. Carbon Cycle Science Plan*, Boulder, Colorado, USA: University Corporation for Atmospheric Research, Action Plan.

Schimel, D., I. Alves, M. Heinman, F. Joos, D. Raynaud and T. Wigley (1996), *Radiative Forcing of Climate Change: CO_2 and the Carbon Cycle*, in IPCC Second Assessment Report on Climate Change: The Science of Climate Change, Cambridge University Press, pp. 76–87.

Schlamadinger, B. and G. Marland (1998), 'The Kyoto Protocol: Provision and Unresolved Issues Relevant to Land-use Change and Forestry', *Environmental Science and Policy*, **1**, 313–27.

Schlamadinger, B. and G. Marland (2000), *Forests, Land Management and the Kyoto Protocol*, Arlington, USA: Pew Center on Global Climate Change, Report Land Use and Global Climate Change project.

UNDP (2000), *World Energy Assessment*, World Energy Council, United Nations Department of Economics and Social Affairs, New York, USA.

UN-ECE/FAO (2000), *Forest Resources of Europe, CIS, North America, Australia, Japan and New Zealand*, United Nations Economic Commission for Europe, Timber and Forest Study Paper 17, Geneva, Switzerland.

Valentini, R., G. Matteucci, A.J. Dolman, E.-D. Schulze, C. Rebmann, E.J. Moors, A. Granier, P. Gross, N.O. Jensen, K. Pilegaard et al. (2000), 'Respiration as the Main Determinant of Carbon Balance in European Forests', *Nature*, **404**, pp. 861–5.

Van Minnen, J.G., K. Klein Goldewijk and R. Leemans (1995), 'The

Importance of Feedback Processes and Vegetation Transition in the Terrestrial Carbon Cycle', *Journal of Biogeography*, **22**, 805–14.

Vrolijk, C., M. Grubb, B. Metz and E. Haites (2000), *Quantifying Kyoto Workshop. A Summary,* London, UK: Royal Institute of International Affairs.

Watson, R.T., I.R. Noble, B. Bolin, N.H. Ravindranth, D. Verado and D.J. Dokken (eds) (2000), *IPCC Special Report on Land Use, Land-Use Change, and Forestry*, Cambridge University Press.

WBGU (1998), *The Accounting of Biological Sinks and Sources under the Kyoto Protocol: A Step Forwards or Backwards for Global Environmental Protection?*, Special Report 3-9806309-0-0, Bremerhaven, Germany.

PART III

Implementation and the Development of a Climate Regime

9. Comprehensive approaches to differentiation of future climate commitments – some options compared

Marcel M. Berk, Joyeeta Gupta and Jaap C. Jansen

1 INTRODUCTION

One of the most contentious issues in international climate policy development is the issue of 'burden sharing' or differentiation of (future) commitments:[1] who should contribute when and how much to mitigate global climate change and to pay for climate change adaptation measures? This issue was particularly debated and analysed in the run-up to the signing of the United Nations Framework Conference on Climate Change (UNFCCC) in Rio de Janeiro in 1992 (see for example Grubb, 1989; Krause et al., 1989; Agarwal and Narain, 1991; Grubb et al., 1992; den Elzen et al., 1992; Grübler and Nakicenovic, 1992; Rose, 1992; Mintzer and Leonard, 1995; Torvanger and Godal, 1999). Eventually, the UNFCCC adopted the provision (Article 3.1) that 'The parties should protect the climate system ... on the basis of *equity* and in accordance with their common but differentiated *responsibilities* and respective *capabilities*' (emphasis is the authors'). It also states that 'Accordingly, the developed country Parties should take the lead' (Article 3.1). It thus divides the world into developed (listed in Annex 1) and developing countries, and indicates clearly that the developed countries should take action first. But it does not provide clarity on what the subsequent steps should be and how the adopted principles should be operationalized.

In 1997, the parties to the Convention agreed on specific quantitative commitments for developed-country parties in the Kyoto Protocol (UNFCCC, 1997). The US, however, made it clear that it demanded 'meaningful participation by key developing countries' as a precondition for its ratification of the new protocol. In 2001, the new Bush administration rejected the Kyoto Protocol as 'unfair' because it does not include quantitative commitments for developing countries, and because implementation would have large negative impacts on the US economy. Initially the rest of the developed world was unwilling to ratify the Protocol without US participation, especially since the US emits at least 25 per cent of global emissions. The outright rejection of the Kyoto Protocol by the US caused the rest of the world to rally together to see what could be undertaken without US participation. This resulted in a political agreement in Bonn during the second part of the Sixth Conference of the Parties (CoP6 bis) and decisions at Marrakesh in November 2001 on further details necessary for implementing the Kyoto Protocol (UNFCCC, 2001a, 2001b). Prospects for ratification of the Kyoto Protocol without US participation are now looking good.

At the same time, the debate about future differentiation of commitments has already started, partly fuelled by the US resistance to the Kyoto Protocol and discussions on voluntary commitments by developing countries under the Protocol.[2] This debate tends to focus on how to involve developing countries in future protocols and identifying what would be acceptable quantitative commitments. However, the issue is much broader. It raises not only questions about a fair distribution of future commitments amongst all countries, both developed and developing, but also questions about the UNFCCC's effectiveness in meeting its ultimate goal.

This chapter aims to explore and evaluate some possible options for future differentiation of commitments as proposed in recent studies. Based on an analysis of the shortcomings of the Kyoto Protocol, it argues in favour of a comprehensive approach that promotes an internally consistent, legitimate regime that builds on defined principles, criteria and rules for differentiating commitments for all countries. This should make the adoption of future commitments predictable, as well as providing certainty about the effectiveness of the regime in meeting the ultimate goal of the Climate Convention: to stabilize global concentrations at safe levels. The chapter describes, compares and evaluates some possible options for such comprehensive approaches, notably the multi-stage, per capita convergence, Triptych and multi-sector convergence (MSC) approaches (see Sections 4 and 5). Finally, it draws some conclusions about the future development of the climate regime (see Section 6).

2   EVALUATION OF KYOTO PROTOCOL COMMITMENTS

During CoP3 in Kyoto in 1997, agreement was reached on binding quantitative emission targets for the developed countries leading to a reduction of CO_2-equivalent emissions of 5.2 per cent in the 2008–2012 period compared to 1990 levels (UNFCCC, 1998). To reach agreement, a substantial differentiation of targets proved necessary. In the Kyoto Protocol of 1997, the European Union (EU), Bulgaria, the Czech Republic, Estonia, Latvia, Romania, Slovakia, Slovenia, Switzerland, Liechtenstein and Monaco had a –8 per cent target, the US a –7 per cent target, Canada, Japan, Hungary, Poland, a –6 per cent target, Croatia a –5 per cent target, the Russian Federation, Ukraine and New Zealand a stabilization target. Norway was given a +1 per cent target, Australia +8 per cent and Iceland +10 per cent.

In terms of differentiation of commitments, the Kyoto Protocol does not set a very good precedent for future negotiations (Berk et al., 2001). First, although there were various proposals for differentiation of commitments among the developed countries from both academic and policy circles (see for example Torvanger et al., 1996; Kawashima, 1996; Blok et al., 1997; Depledge, 2000), there is no easily discernible logic to the differentiation of the targets for the developed countries. One might argue that Japan, already very efficient, has less room to reduce its emissions than many other developed countries. For the rest, the eventual outcome of the negotiations seems to have been the result of hard bargaining, and determined primarily by what the various parties were willing to commit themselves to. The EU first promoted a flat 15 per cent reduction target, but its claim for internal differentiation (allowing some EU member states like Portugal and Greece to substantially increase their emissions, to be compensated by stronger reductions by other members) weakened its position and led some other developed countries (like Australia) to claim the right to increase their emissions as well, much to the annoyance of the developing countries. At the same time, the Ukraine and Russia were allowed stabilization targets, in spite of the fact that their current and future (2008–2012) emission levels are likely to be considerably lower than their 1990 emissions. Combined with the fact that the Kyoto Protocol allows countries to trade emission reductions, this has created the problem of 'hot air' trading which substantially reduces the effectiveness of the KP (see den Elzen and de Moor, 2001). In order to secure the participation of all developed countries, the negotiations have resulted in a situation where countries that bargained hard got exceptional allowances, while others committed themselves to lower targets than they were originally willing to accept. Without accepted principles and rules for determining a fair differentiation of commitments, therefore, negotiations resulted in a watering-down of the overall emission reduction target, and weakened the principle that developed countries should take the lead by reducing their emissions.

Second, the Kyoto Protocol left the 'rules of the game' unsettled. These rules concerned many issues affecting the stringency and costs of the agreed targets, such as the use of the Kyoto mechanisms and sinks options in meeting emission targets. In the post-Kyoto negotiations, these technical issues have been used to 're-negotiate' the commitments. After the US rejected the Kyoto Protocol, Japan, Canada, Australia and Russia exerted effective pressure on the EU and the 'G77 and China' to accept decisions during CoP6bis in Bonn and CoP7 in Marrakesh that includes no quantified restrictions on the use of the Kyoto mechanisms[3] and allow for extensive use of sink options. These decisions substantially affect the stringency of both the overall commitments and those of individual countries (particularly Canada and Japan) and are likely to also affect the development of the Kyoto mechanisms.[4]

Third, the Kyoto Protocol is based on a simple division between developed and developing countries that will be problematic in future climate change negotiations. Many developed countries resent the fact that countries like Singapore, South Korea and Mexico have not adopted quantitative commitments, while Turkey has refused to take on commitments under the Kyoto Protocol and has requested that it be removed from the Annex I list because it is not a high-income country. The internal differentiation arrangement within the EU has led other developed and developing countries to accuse the EU of inconsistency between internal and external EU climate policies (Gupta and Van der Grijp, 1999). This indicates that within the climate change regime, the EU cannot treat internal and external rules for differentiation of commitments independently, but that such rules need to be based on more generally shared principles.

Fourth, the way the Kyoto Protocol was negotiated has been criticized by many developing countries. Much of the negotiating took place in informal and parallel sessions, which limited the access of developing countries to the negotiation process and its transparency. The final Protocol was also adopted after many of the developing country negotiators had already left for their home countries. This has led the G77 and China to operate as a closed and defensive bloc, notwithstanding their internal differences of opinion and interests. For negotiations on future differentiation of commitments, a more open and transparent negotiation process that allows for more equal participation by developing countries will be important to overcome the present divide between the blocs (see Mwandosya, 1999 and Gupta, 2000).

Finally, the Kyoto Protocol takes a short-term approach to commitments, with no long-term perspective. This encourages short-term actions to meet the quantitative Kyoto target that may be incompatible with requirements for stabilizing greenhouse gas concentrations in the long term.

3   COMPREHENSIVE APPROACH

What are the implications of the experience with differentiation of commitments during the Kyoto Protocol negotiations for the future of the climate regime? The lack of a systematic approach or policy architecture (Jacoby et al., 1999) in the way the Kyoto targets were determined poses a potential threat to the development of an effective, and for all countries acceptable, future climate regime. An incremental evolution of the climate regime in the form of a gradual extension of the Annex I group on the basis of voluntary participation and ad hoc definitions of quantitative commitments is unlikely to bring about the level of global emission control needed to keep open the option of stabilizing CO_2 concentrations at 450 ppmv (or 550 ppmv including all gases) (Berk et al., 2001). Not only will this result in uncertainty about the environmental effectiveness of the future climate regime, it will encourage countries to free-ride. This is likely to have a negative impact on public support for climate policies in countries already taking on quantified targets. At the same time, it is unlikely to constructively engage developing countries into taking on new, quantitative commitments by overcoming their present distrust. Finally, it will also result in uncertainty about climate policies for the private sector, which hinders long-term investment planning.

There is thus a need to develop a *comprehensive approach*, which is defined as an international climate regime that contains guiding principles, criteria and rules for differentiating future commitments for all parties. The principles, criteria and rules should be thought of as providing default positions subject to further negotiations (for example on exemptions, and so on). This will increase the transparency, legitimacy and predictability of the regime, which will make it possible for countries and industries to plan their climate policies accordingly.

4   OPTIONS FOR COMPREHENSIVE APPROACHES

4.1   Multi-stage Approach

One option for a comprehensive approach is to adopt a multi-stage approach (Gupta, 1999; Gupta et al., 2001; den Elzen et al., 2001). This approach consists of a system to divide countries into groups with different levels of responsibility or commitments (stages). The aim of such a system is to ensure that countries with comparable circumstances in economic, developmental and environmental terms have comparable responsibilities/commitments under the climate regime. Moreover, as their circumstances change, the system defines when their level of responsibility/commitment changes.

In the context of such a system, Gupta et al. (2001) use country criteria and so-called 'graduation profiles'.[5] Their graduation profile system is based on easily available comparable data, namely per capita income levels (financial ability) and CO_2 from industrial emissions (responsibility for causing the problem). Using a combination of these two criteria, 12 potential categories of countries have been identified (see Table 9.1). The criteria used are merely indicative of the kinds of criteria that can be used. The Human Development Index (HDI) is also used as an indicator of the ability of countries to take action, and their vulnerability to climate change. The method consists of an inclusion level (countries below a certain level for one or more given criteria are automatically included in a specific group) and a graduation level (countries above a certain level for one or more given criteria for a specified period of time, (for example three years) are automatically included in the next group after two consecutive periods of graduation. These countries are 'on notice' with regard to taking on the obligations that the next group is already subject to. A list of quantitative and qualitative obligations was developed, and tailor-made packages of responsibilities were defined for countries falling into each group (see Gupta, 1999; Gupta et al., 2001). Further, the system could allow for specific exceptions for countries with special domestic circumstances.

A quantitative and, compared to Gupta et al. (2001), simplified version of the multi-stage approach can be found in the FAIR (Framework to Assess International Regimes for differentiating of commitments) model as part of the so-called 'increasing participation' mode (den Elzen et al., 2001). In this model the number of parties involved and the level of involvement in global emission control gradually increases over time according to participation and burden-sharing rules. These rules are not fixed, but can be selected from a range of options, including per capita income levels, (per capita) emission levels or the contribution to global warming (Brazilian Proposal) (UNFCC,

Table 9.1 Criteria for classifying countries and packages of obligations

	Emissions below minimum per capita	Emissions in the middle category	Emissions above maximum per capita
Very high income per capita	Package 1c Including, inter alia: • quantitative emission stabilization commitments • voluntary contribution to adaptation fund	Package 1b Including, inter alia: • quantitative commitments for interim emission stabilization followed by further reduction • compulsory contribution to adaptation fund • non-compliance fine	Package 1a Including, inter alia: • quantitative emission reduction commitments • compulsory contribution to adaptation fund • non-compliance fine
High income per capita	Package 2c Including, inter alia: • quantitative emission limitation commitments • voluntary contribution to adaptation fund	Package 2b Including, inter alia: • quantitative commitments for interim emission stabilization followed by further reduction • voluntary contribution to adaptation fund	Package 1b/2a Same as package 1b
Medium income per capita	Package 3c Including, inter alia: • listing as priority group 2 for support from adaptation fund • receive permits for emissions below minimum per capita	Package 3b Including, inter alia: • listing as priority group 3 for support from the adaptation fund • policies and measures	Package 3a Including, inter alia: • listing as priority group 4 for support from the adaptation fund • policies and measures

Table 9.1 continued

	Emissions below minimum per capita	Emissions in the middle category	Emissions above maximum per capita
Low income per capita	Package 4c Including, inter alia: • listing as priority group 1 for support from adaptation fund • receive permits for emissions below minimum per capita	Package 4b Including, inter alia: • listing as priority group 2 for support from the adaptation fund	Package 4a Including, inter alia: • listing as priority group 3 for support from adaptation fund • voluntary policies and measures

1998). The increasing participation approach was first introduced at the fourth Conference of the Parties (CoP-4) as part of an evaluation of the Brazilian proposal (Berk and den Elzen, 1998; den Elzen et al., 1999). The approach was later extended with more stages into a multi-stage approach which was presented during CoP6 (IISD, 2000; den Elzen et al., 2001).

The increasing participation mode now offers a regime with four stages in all to differentiate commitments among parties over time. It can be summarized as follows:

Stage 1 Reference scenario
Non-participating parties (non-Annex I) first follow their baseline emission scenario (reference scenario) until they meet a threshold for taking on decarbonization targets. Annex 1 countries participate from the start (for example with the Kyoto Protocol commitments).

Stage 2 Decarbonization targets
The parties then enter a stage in which their allowable emissions are controlled by decarbonization targets, defined by the rate of reduction in the carbon intensity of their economy. A region leaves this stage when it attains any of the selected participation thresholds for burdensharing.

Stage 3 Stabilization of emissions
The parties first enter an emissions stabilization period, in which they stabilize their absolute or per capita emissions for a user-defined number of years before actually entering the burden-sharing regime.

Stage 4 Emission reduction regime
Emission reduction or burden-sharing rules determine the contribution of each participating region to the overall emission reduction needed to meet the target emission scenario.

Ideally, the decarbonization stage results in a reduction of the increase in allowable emissions. The stabilization stage then acts as an intermediate stage between an increase and subsequent decrease in allowable emissions. Quantitative results of analyses of the multi-stage approach in FAIR have been presented elsewhere (den Elzen et al., 2001; Berk and den Elzen, 2001).

4.2 Per Capita Convergence Approach

In the per capita convergence approach all parties immediately participate in the emission control regime (after the first commitment period), with per capita emission rights/permits converging towards equal levels over time. The Global Commons Institute (GCI) first introduced the 'Contraction and Convergence' approach. Early results of the approach were published to good effect at COP-2 and have been distributed widely since then. The procedure is as follows. First, a global emissions profile or global greenhouse gas emissions contraction budget is defined which may be linked to a long-term atmospheric greenhouse gas concentration target, for example 450 ppmv CO_2. This budget is then allocated to countries in such a way that their per capita emissions converge from their present diverse values to a global average world value which is the same for each country in the convergence year (Meyer, 2000).

The convergence of per capita emissions can take place in either a linear or non-linear way. In the linear approach, countries' shares in the yearly global emission budget converge linearly from their shares in global emissions in the starting year to an allocation based on their shares in global population in the convergence year. In the case of non-linear convergence, as developed by the GCI, the convergence in per capita emissions takes place either mainly in the first part or the last part of the convergence period, depending on the convergence rate chosen. The Indian Centre of Science and Environment (CSE) suggested a special variant of per capita convergence in which the concept is combined with basic sustainable emission rights (Agarwal et al., 1999; CSE, 1998; Agarwal, 1999). The approach is related to both the idea of 'survival emissions' as well as to the idea of commons as natural sink for CO_2 (in particular the oceans). It starts from the concept of basic emission rights per person, which are linked to a so-called global 'sustainable' emission level. This level is determined by the amount of CO_2 that can be emitted into the atmosphere without increasing the atmospheric concentration of CO_2. Before applying the convergence approach, this 'sustainable' part of the global emission budget is first distributed on a per capita basis. Besides this basic emission allowance, each person has a remaining emission allowance, which is calculated using the convergence methodology (as described above), but now using the profile of the remaining global emissions. This remaining global emissions profile is determined by the global target emission profile minus the global 'sustainable' emission level. All three versions of per capita convergence have been incorporated in the FAIR model. The main policy variables in the per capita convergence approach are:

- the level of contraction of global emissions: this not only affects the level of climate protection but also the distribution of cumulative emission permits/rights during the convergence period. The stronger the contraction, the smaller the share of developing countries in the cumulative emissions.
- the convergence year: the shorter the transition period, the larger the share of developing countries in the cumulative emissions, and the steeper the reductions of developed country permits/rights over time.
- the rate of convergence: the higher the rate, the larger the share of developing countries in the cumulative emissions, and the steeper the reductions of developed country permits/rights over time.
- the level of sustainable emissions (in the CSE variant).
- population (in all variants): in the calculation of emissions permits/rights there is the option to set a cap on accounting for population growth in order not to 'reward' future population growth.

While the type of convergence and the convergence year (and the population 'cut-off year') probably need to be fixed from the start, the global greenhouse gas emissions budget for each subsequent commitment period can be the subject of continuous negotiation (taking into account new information about the severity of climate impacts and costs of mitigation).

4.3 The Triptych Approach

The Triptych approach to differentiation (Phylipsen et al., 1998; Groenenberg et al., 2001) is a sector-oriented approach that was used to support decision-making in the EU prior to Kyoto (COP-3) (Blok et al., 1997). It has subsequently been adapted for use at the global level (Groenenberg et al., 2001). In contrast to the top-down per capita convergence approach (from global emission ceilings to regional emission budgets), the Triptych approach is more bottom-up in nature, although it can be combined with specific emission targets (as illustrated in the case of the EU). Up to now, the Triptych approach has only set targets for CO_2; other greenhouse gases might be included in a future version (Phylipsen et al., 1998). It is one of the few approaches to have taken into account different national circumstances, such as differences in standard of living, fuel mix, economic structure and the competitiveness of internationally-oriented industry. The name Triptych alludes to the distinction between three emission sectors:

- *Domestic sectors.* These comprise the residential, commercial, transportation, light industry and agricultural sectors, where emission reductions can be achieved by national measures. Emissions are taken to converge on a per capita basis in a certain convergence year. The application of an assumed reduction rate to the per capita base-year level throughout the prescribed convergence period gives the convergence level. This convergence rule is applied to all countries, which tends to work out quite generously for poor developing countries with very low emission levels.

- *Internationally oriented energy-intensive industry.* This sector comprises internationally-oriented manufacturing (sub)sectors, where competitiveness is determined by the costs of energy and energy-efficiency measures. These are the building materials industry, the chemical, iron and steel industry, non-ferrous metals, pulp and paper industries, refineries, coke ovens, gasworks and other energy-transformation industries (excluding electricity generation). The development of the emissions in heavy industry is assumed to depend on the growth of the physical production rather than on the growth of industrial value added, as emissions are linked more closely to the production level than to the valuation of industrial output on the market. In the global study (Groenenberg et al., 2001), projections of future growth rates for the physical production in the various sub-sectors of the energy-intensive industry are differentiated for OECD countries, economies in transition and developing countries. As physical production is the point of departure, energy efficiency expressed in physical terms is considered rather than the energy intensity of the economy expressed in monetary terms. Energy efficiency in the global study is combined with decarbonization into one rate, which in this assessment is set at 1 per cent for all countries. It is argued that setting CO_2 emission targets on a per capita basis would be to the disadvantage of the competitiveness of industries in countries with a high share of energy-intensive manufacturing sub-sectors. Separate rules for energy-intensive industry would make special allowance for these considerations, to avoid biases in international trade.

- *Electricity-generation sector.* There are large differences between regions and countries in the share of power production techniques like nuclear power and renewables and in the (fossil) fuel mix in this sector. The potential for renewable energy is different for each region, as is the public acceptance of nuclear energy. Emission targets for this sector are based on projections of national electricity consumption. These projections, in turn, are based on projections of regional economic growth. No intra-regional differences in economic growth among countries are assumed to occur. Furthermore it is assumed that efficiency in end-use appliances will reduce demand by a certain percentage each year. In addition, specific criteria with respect to fuel mix in 2015 have been formulated for the numerical elaboration for 2015, most of which are similar to those applied in the Triptych study for the European Union (Phylipsen, et al., 1998). These concern an assumed reduction of both coal- and oil-based electricity generation, a constant share of nuclear power compared to 1990 as an approximation of national preferences, a targeted increase in the share of renewables and Combined Heat and Power (CHP) generation and the generation of the remainder of the electricity demand based on natural gas. Efficiencies of generation capacity for solid, liquid and gaseous fossil fuels are assumed to converge by the year 2030.

National emission allowances are calculated by adding up the emission allowances resulting from applying the specific rules to each of these sectors. The Triptych approach requires a baseline scenario for CO_2 emissions in order to establish future allowances, as well as population projections and projections of economic growth and physical production in energy-intensive industries. A more extensive description of the (original) methodology and its background can be found in Phylipsen et al. (1998).

4.4 Multi-Sector Convergence Approach

The Multi-Sector Convergence (MSC) approach sets out to derive comprehensive greenhouse gas emission targets for commitment periods after the first one defined by the Kyoto Protocol (that is 2008–2012) which can be readily rolled over from one to the other, allowing for periodic re-calibration of parameter values. The design of the approach is based on an extensive evaluation of historical proposals for burden differentiation of greenhouse gas abatement (Ringius et al., 1999). The philosophy behind the approach is to account for principles of responsibility, capacity and need (see Section 5).

and to be comprehensive and flexible. The fact that climatic change is viewed as a common, potential threat to the existence of humankind in both industrialized and developing countries makes convergence approaches attractive in principle. Yet, at the same time, the approach tries to account for the large differences in national circumstances that make for variations of per capita emission requirements among countries, and demand flexibility in negotiating targets.

Paramount considerations for introducing flexibility are:

- *Level and time-path of allowable global greenhouse gas emission levels.* Given the large uncertainties, ample negotiation space for convergence year(s) and convergence level(s) should be built in.
- *Allowance for low-emission countries.* The MSC approach postulates that countries with per capita emissions higher than a certain global allowable emission standard have to take on the cost burden of mitigation efforts, whereas so-called 'low-emission countries' should have the right to economic development without incurring the additional cost of emission mitigation. Once the per capita allowable global emission level is reached, MSC prescribes that the signatory countries concerned should be called on to accept national emission mitigation targets.
- *Adjustment for rigidities and other major concerns at the sector level.* Distinct structures of the national economies constitute a significant factor in determining the possibilities of the adjustment process in terms of mitigation burden and speed. Some sectors have a slower turnover rate of greenhouse gas-emitting infrastructures than others. Furthermore, certain sectors might be considered to be providing more essential services for the fulfillment of basic human needs than others. Hence, an approach marrying top-down features to ensure environmental integrity, with bottom-up features to allow for national circumstances, appears to be warranted.
- *Allowance for constant location-specific factors.* Per capita emissions in areas with extremely cold or hot climates will be appreciably higher than those in temperate climates, all other factors being equal. The same goes for sparsely populated areas compared to densely populated areas. Also, in terms of renewable energy potential that can be harnessed in an economically feasible way, the resource base significantly affects the cost of low-emission energy.
- *Allowance for changeable location-specific factors.* Some countries are faced with severe economic transition problems. One category of countries is that of the high-emission countries with a strongly centralized economy moving towards a market-oriented economic

system. Another category comprises high-emission countries with an economy that is presently very dependent on the export of fossil fuels. These countries may negotiate assigned emission levels that are temporarily equal to, or not very far below, their emission baselines.

Distinction between different sectors

In the MSC approach, the sector categorization of greenhouse gas emissions is based on a number of considerations:

- the significance of the sector's contribution to global greenhouse gas emissions;
- the sector should not be overly heterogeneous with respect to intra-sector greenhouse gas-emitting characteristics;
- acceptable and reliable sector data, notably on greenhouse gas emissions, should be available;
- the sectors should be as comprehensive as possible with respect to greenhouse gas emissions (allowing for the aforementioned considerations), but their number should be small enough to preserve simplicity and clarity.

Given these considerations, the following sector categorization is proposed for the short term:

1. *Power*: Greenhouse gas emissions related to power generation, excluding power use in heavy industry.
2. *Households*: Non-power-related greenhouse gas emissions.
3. *Transportation*: Non-power-related greenhouse gas emissions.
4. *Heavy industry*: Greenhouse gas emissions from energy use (including power) and process emissions (especially in the cement industry).
5. *Services*: Non-power-related greenhouse gas emissions from other economic activities, mainly services.
6. *Agriculture*: Non-power-related greenhouse gas emissions from crop production (mainly methane from paddy-growing) and livestock (notably methane, some nitrous oxide).
7. *Waste*: Non-power-related greenhouse gas emissions from waste treatment (mainly from landfills: especially methane, some nitrous oxide).

Together these sectors account for the vast majority of greenhouse gases covered by the Kyoto Protocol. The main exceptions are greenhouse gas emissions due to land-use changes, and emissions of HFCs, SF_6 and PFCs. With the application of MSC, the sector categorization can be suitably adjusted following developments on the diplomatic front (such as the inclusion of LULUCF) and once improved data has been made available.

Global sector emission norms

Points of departure are (a) the global per capita emission levels in the base year at aggregate and sector level, (b) the (negotiated) convergence year, and (c) the (negotiated) *global* per capita total emission standard (PCTES) in the convergence year. The *global* per capita sector emission standards (PCSESs) in the convergence year are proposed by allowing for sector-specific characteristics, such as projected developments with respect to driving factors with regard to sector emission levels and inertia in sector infrastructures. Corresponding standards in intervening years are obtained through geometric interpolation.

National emission mitigation targets

The starting point for each country under the mitigation commitment regime are the actual per capita sector emission levels in the base year. In the years thereafter, these levels are supposed to converge to the global sector emission standards of the convergence year. Non-binding *national* PCSESs in intermediate target years are obtained by geometric interpolation between the national sector emission levels in the base year and *global* PCSESs of the convergence year. The *national* PCSESs are added up and multiplied by the total population in order to determine basic national emission mitigation targets for the countries and years concerned.

Inclusion of allowance factors

The previous three steps delineate the basics of the MSC framework. Yet, the framework is made more flexible by the inclusion of so-called country-specific allowance factors, which, all other factors being the same, can significantly influence differences in greenhouse gas emission requirements between countries. Which allowance factors should be taken into account, and how, is up to the actors at the negotiating table. Examples of possible allowance factors are:

- *Population density.* Thinly populated countries might be allotted additional per capita emission allowances inversely proportional to their population density.
- *Climate.* Countries might be allotted additional per capita emission assignments in line with their number of heating or cooling-degree days.
- *Agriculture.* Rice cultivation is the most important source of greenhouse gas emissions (methane) in the food crop sector. As rice tends to meet the

basic nutritional needs of a large part of the world population, part of the
per capita standard emission mitigation amounts related to this activity
might be forfeited.
* *Economic transition problems.* More lenient national emission targets
  might be conceded temporarily to countries facing transitional problems,
  including fossil-fuel exporters.

In order to ensure compliance with those global emission goals that receive
the necessary backing during international negotiations, the MSC approach
should be applied iteratively. A more extensive description of the approach
and numerical examples can be found in Jansen et al. (2001a and 2001b) and
Sijm et al. (2001).

5   COMPARISON AND EVALUATION OF APPROACHES

In this section we will compare the various approaches, and make a
preliminary evaluation of possible strengths and weaknesses of the various
approaches. First, we will compare the approaches by looking at a number of
different dimensions of climate regimes. Second we will use certain criteria
to evaluate the various approaches.

5.1   Equity Principles

Normative principles for equity or distributional fairness are a key element of
any comprehensive regime for the differentiation of future mitigation
commitments. There is no commonly accepted definition of equity or
distributional fairness (see for example Rose, 1992; Banuri et al., 1996;
Ringius et al., 1999). To compare the four comprehensive approaches we use
the rather simple typology developed by Ringius et al. (1999), which
distinguishes three relevant of principles for distributive fairness:

* *Responsibility:* costs should be distributed in proportion to a party's share
  of responsibility for causing the problem.
* *Capacity:* costs should be distributed in proportion to ability to pay.
* *Need:* all individuals have equal rights to pollution permits with as the
  bottom line the right to basic human rights, including a decent standard
  of living.

In international environmental negotiations, the default option is usually the
rule of equal obligations (for example flat-rate emission reductions). This rule
is based on the principle of sovereignty in international law, which implies

that all states are equal (and thus have equal rights and responsibilities) and have an exclusive right to govern their own territory. In international environmental negotiations, the principle is also used to claim status quo rights. The principle is omitted in the typology of Ringius et al. (1999) because it no longer seems to have the same legitimacy in international environmental negotiations as the other principles.[6]

In addition to equity principles, there are a number of other dimensions of the architecture of possible regimes for the differentiation of future commitments (den Elzen et al., 2001; Berk et al., 2001).

5.2 Problem Definition: Burden-sharing or Resource-sharing

The climate change problem can be defined as a pollution problem or as a global commons issue. These different concepts have implications for the design of climate change regimes. In the first concept, the issue is burden-sharing, and the focus will be on defining who should reduce or limit their pollution and by how much. In the latter concept, the issue is resource-sharing, and the focus is on who has what user rights. Mitigation efforts will be determined by the distribution of user rights.[7]

5.3 Overall Emission Limit or Not?

One can define the emission reduction top-down by first defining globally-allowed emissions, and then apply certain participation and differentiation rules for allocating the overall reduction effort needed. Alternatively, emission control efforts among parties can be allocated bottom-up, without a predefined overall emission reduction effort. In the top-down approach, the question of adequacy of commitments is separated from the issue of differentiation of commitments. In the bottom-up approach, the two are usually dealt with at the same time. The stringency of country and overall targets are determined in an iterative way.

5.4 Participation

Another dimension is the degree of participation: who should participate in mitigation and when does their obligation begin? This issue relates to both the types of thresholds (for example per capita income or emissions) for participation, and the threshold level or the timing. At the same time, there is no need for all parties to participate in the same way. This points to yet another dimension: the form of the commitment.

5.5   Form of Commitment

The form of the commitment for countries can be equal for all, like the binding emission targets in the Kyoto Protocol, but they could also be defined in differentiated ways (see, for example Claussen and McNeilly, 1998; Baumert et al., 1999; Philibert and Pershing, 2001). Instead of absolute or fixed targets, commitments may be defined as relative or dynamic targets, such as a reduction in energy and/or carbon intensity levels. There is also the option of non-binding targets. Finally, commitments could also consist of sector targets, or common policies and measures.

We can use these dimensions to identify important differences between the various approaches presented (Table 9.2). The per capita convergence, MSC and Triptych approaches do not really account for the responsibility principle – only the multi-stage approach does. The per capita convergence approach is the only approach based on the resource sharing problem definition. While adopting some elements of the per capita convergence approach, the Triptych approach defines the climate problem as a pollution issue. Although the MSC approach takes the global commons issue as its point of departure, the elaboration of this approach has pollution-problem-definition aspects as well, contingent on the role allotted to allowance factors in its operationalization.

The per capita convergence approach is most clearly top-down in character. This is also the case for the multi-stage approach as implemented in the FAIR model, but not necessarily so as shown by Gupta et al. (2001; Gupta, 1999). Since all approaches are comprehensive in nature all assume full participation in the sense that all countries are part of the regime, but there can be thresholds that in fact exempt them from having to take action yet. In the per capita convergence, MSC and Triptych approach the form of the commitment will be the same for all countries (emission permits), apart from possible exemptions. Only the multi-stage approach incorporates different forms of commitments (for example decarbonization targets in addition to emission targets).

Table 9.2 Comparison of approaches to international burden differentiation

Dimensions	Multi-stage	Per capita convergence	Global Triptych	Multi-sector convergence
Equity principles				
• Responsibility	X			
• Capability	X	(X)	X	X
• Need	X	X	X	X
Problem definition				
• Pollution problem	X		X	(X)
• Global commons issue		X		X
Emissions limit				
• Top down	X *	X		(X)
• Bottom up / iterative	(X)**		X	X
Participation				
• Partial	(X)			(X)
• All	X	X	X	X
Form of Commitment				
• Same		X	X	X
• Different	X			

Notes:
X = applicable.
(X) = partly applicable.
* the multi-stage in FAIR model.
** multi-stage.

Source: According to Gupta et al. (2001).

5.6   Strengths and Weaknesses of the Approaches

We will evaluate the various approaches on the basis of the following criteria (Berk et al., 2001; Torvanger and Ringius, 2000):

* comprehensiveness regarding equity principles;
* environmental effectiveness;
* economic efficiency;
* flexibility;
* simplicity and operational requirements.

5.7 Comprehensiveness Regarding Equity Principles

As indicated in Table 9.2 the multi-stage, MSC and Triptych approaches are based on more than one equity principle. The convergence approach is mainly based on the need principle, but to some extent also on capability by allowing for a transition period.

5.8 Environmental Effectiveness

Environmental effectiveness is defined as the overall global level of emission control that the climate regime is able to provide. This in turn is affected by various factors, such as the stringency of commitments adopted, the level of certainty about emissions reductions and 'leakage' (the offset of emission reductions in one country by increases in other countries). In principle, the best way to ensure environmental effectiveness is if a climate regime is based on pre-set global emission targets and commitments on fixed quantitative emission limitations, with all countries participating (provided sufficient compliance can be achieved). Without full participation, there will be leakage to countries without emission limitations. Carbon leakage may also occur in the case of countries with relative targets since they also have no absolute emission cap.

In principle, a top-down approach such as the per capita convergence approach with a global emission cap seems more likely to ensure environmental effectiveness than the other approaches. However, the negotiation of the overall global target is likely to be affected by an assessment of the costs and efforts involved. It can be argued that the level of flexibility in target differentiation and the cost-effectiveness of the regime could be more important for the stringency of the target adopted than following a top-down approach.[8] Conversely, bottom-up approaches are likely to be more susceptible to lobbying by sector groups.

In the multi-stage approach (and to a lesser extent in the MSC approach) the environmental effectiveness is less certain than in the other approaches because of uncertainty about baseline emissions of non-participating countries and the risk of substantial carbon leakage. All regimes allow for *ex-post* evaluation of the environmental effectiveness after each commitment period, but since commitments are usually decided upon well before the start of commitment periods there will be a substantial time delay.

5.9 Economic Efficiency

The adoption of the Kyoto mechanisms has drastically changed the setting for discussing the economic efficiency of various regimes for differentiating commitments. In principle, their introduction has offered the possibility of attaining a high level of economic efficiency regardless of the way mitigation efforts are distributed. Via the Clean Development Mechanism (CDM), emission reduction options in countries without emission targets can also be used to enhance the efficiency of greenhouse gas abatement, albeit at generally higher transaction costs due to its project-based character. Some authors have pointed out that regimes allowing developing country parties to take on relative or dynamic targets, like decarbonization targets in the multi-stage approach, may also allow them to join in emission trading to the extent that their improvements exceed their targets (Hargrave, 1998; Baumert et al., 1998; Philibert, 2000; Philibert and Pershing, 2001). Others argue that such targets will complicate the functioning of emission trading systems (Baumert et al., 1998; Müller et al., 2001).

Compared to other approaches, the per capita convergence and a global Triptych approach seem to offer the best opportunities to fully explore cost-reducing options because all parties can fully participate in global emission trading. At the same time, a per capita convergence approach and a global Triptych may result in excess emission allowances for developing countries (depending on baselines and the stringency of global emission targets). However, this will not affect the efficiency of the regime, only the distribution of costs.

5.10 Flexibility

Flexibility of a regime approach is important for two reasons. First, a regime should be able to take into account the special circumstances of particular (groups of) countries that require exceptions or adjustments to the application of general rules, for instance for reasons of fairness or political feasibility. Second, the regime should be flexible enough to adapt to changing circumstances created by new scientific information. While the Kyoto mechanisms have reduced the need for flexibility in determining national commitments, flexibility in allocating emissions allowances is still important because it affects the distribution of costs.

In principle, the bottom-up Triptych and MCS approaches offer the most flexibility, while the top-down per capita convergence approach offers the least. However, in practice the convergence approach could be made flexible by adding correction factors to the overall per capita distribution of emission rights to account for special circumstances. Triptych and MSC offer distinct opportunities to account for sectoral differences between countries. To this end, the MSC approach distinguishes several sectors and includes a system of allowance factors for further refinement. The Triptych approach is coarser, but does account for country-specific baseline projections. The flexibility of the multi-stage approach is mainly based on a differentiation of the type of commitments during the various stages, and the selection of thresholds for switching from one stage to the other. However, once a country enters the system, there is little flexibility, although correction factors could be introduced here as well.

5.11 Simplicity and Operational Requirements

Simplicity is an important feature of any future climate regime. A complex regime design creates three substantive problems: it becomes hard for policymakers to assess its fairness and its implications, policymakers may have to rely on complex non-transparent models both to determine targets and to evaluate their implications, and monitoring of compliance may become difficult. Complex regimes also create communication problems, because politicians have to explain and 'sell' the regime to their national constituencies.

Of the various approaches, the per capita convergence approach is clearly the simplest and also the easiest to communicate. The complexity of the multi-stage approach is dependent on the participation and burden-sharing rules chosen: accounting for historical contributions to climate change will introduce large data uncertainties (historical emissions), as well as the need for complex modelling frameworks (as in the Brazilian proposal). A serious complication in both the Triptych and multi-stage approach is the need to determine baseline emissions for target-setting. Both the Triptych and MSC approach require detailed sector information. This requirement is not necessarily an important obstacle for policymakers. The advantage of bottom-up approaches is that they provide policymakers with a better insight into the efforts required, and the fairness of the distribution of efforts among all parties. This has been a major factor in the successful application of the Triptych approach in the EU.

With respect to operational requirements it is obvious that any comprehensive regime that does define quantified commitments for less-developed countries with poor statistical registration and verification systems, will run into substantial operational problems with respect to monitoring of compliance. This will be less a problem in approaches that exempt the least-developed countries from any quantified commitments, like the multi-stage and MSC approaches. However, it is conceivable that in a global Triptych approach, the least-developed countries would also be exempted for this reason.

Table 9.3   Evaluation of different approaches to comprehensive climate change regimes

Dimensions	Multi-stage	Convergence	Global Triptych	Multi-sector convergence
Coverage of different equity principles	++	−	+	+
Environmental effectiveness	+/−*	++	+/−	+/−
Economic efficiency**	+	++	++	+
Level of flexibility	+/−	——	+	++
Level of simplicity/ ease of implementation	+/−	++	−	−

Notes:
++ = very good.
+ = good.
+/− = fair.
− = moderate.
—— = poor.
*: ++ when combined with a global emission ceiling as in FAIR.
**: assuming the use of the Kyoto mechanisms.

6   CONCLUSIONS

This chapter has explored and evaluated some possible options for future differentiation of commitments for developed and developing countries, taking into account the experience with the Kyoto Protocol negotiations (see Table 9.3). The Kyoto outcome seems to have been mainly determined by what the various parties were willing to do, and by an effort to secure the participation of all developed countries. If such an opportunistic approach were to be continued in the future, it seems unlikely to be effective in keeping open the option of stabilizing greenhouse gas concentrations at low levels, and in overcoming the present distrust hindering constructive engagement of developing countries in taking on quantitative commitments.[9]

Based on this analysis, this chapter has argued in favour of a comprehensive approach, namely a regime that is internally consistent and legitimate, and that includes principles, criteria and rules for differentiating commitments for all countries. By presenting four different examples – the multi-stage, per capita convergence, Triptych and Multi-Sector Convergence approaches – it has shown there are many options available.

By way of a preliminary assessment of the strengths and weaknesses, the various comprehensive approaches were evaluated on the basis of a number of criteria: comprehensiveness regarding equity principles; environmental effectiveness; economic efficiency; flexibility and simplicity, and operational requirements. The strengths of the multi-stage, MSC and Global Triptych approaches are that they, unlike the per capita convergence approach, cover more than one equity principle and provide for flexibility in allocating emission allowances, although in different ways. The strengths of the per capita convergence approach are its simplicity, its environmental effectiveness, its optimal level of economic efficiency, and its attractiveness for many developing countries. It weaknesses are its lack of flexibility, and the fact that its roots are in the politically controversial resource-sharing paradigm.

Negotiating a comprehensive approach will be time-consuming and difficult. Although not a good model for future regimes, the Kyoto Protocol provides us with the opportunity to discuss and negotiate a better follow-up, while we can learn how to design and operate those new mechanisms that seem to be valuable to all future approaches.

NOTES

1. Although 'burden-sharing' is a common concept in the literature, this debate is likely to be (re)framed in terms of 'differentiation of (future) commitments' given the language in the UNFCCC. Moreover, the term 'burden-sharing' has a negative connotation and is linked to the framing of the issue as a pollution problem, which is not generally accepted (see Section 5).
2. During CoP3 in 1997 and CoP4 in 1998, considerable pressure was exerted on developing countries to adopt new commitments. This was strongly resisted by the developing countries, but Argentina and Kazakhstan agreed to take on commitments voluntarily, although these have not yet been formalized.
3. Except to prevent overselling, and by limiting the share of credits from afforestation and reforestation projects under the CDM (UNFCCC, 2001a).
4. Without US participation, the lack of restrictions on emission trading has made the problem of hot air a very serious problem. The supply of hot air may outweigh demand for emission credits. Without restriction of hot air supply, this may result in a very sluggish market with low prices. The compromises on the use of the sinks have enhanced this problem further. See for a detailed analysis den Elzen and de Moor (2001).

5. The term 'graduation profiles' is used in UN circles to define the movement of countries from one group to another for the purpose of specific obligations or benefits.
6. The principle of (absolute) sovereignty has been eroded substantially in international law: countries have the right to govern their own territory so long as they do not cause harm to other states. It therefore no longer legitimizes unlimited greenhouse gas emissions. However, it has been pointed out that it is not necessary to resort to the sovereignty principle in international law to legitimize a claim of historical rights or entitlements. They can be legitimately argued for on grounds of equity considerations independent of the sovereignty principle. The claim of status quo rights and related proposals for a flat-rate reduction or grandfathering of greenhouse gas emission permits thus still have some relevance and cannot be easily dismissed (Müller, 1999).
7. Alternatively, the climate change problem can be viewed as part of our unsustainable development path, and its solution an integral part of sustainable development policies. In that case, the focus would no longer be on the development of a climate regime and defining climate targets, but instead the focus will shift to the development and distribution of sustainable technologies and lifestyles. This goes beyond the options presented here.
8. See also Philibert and Pershing (2001).
9. The US withdrawal from the climate change regime has unfortunately only served to exacerbate the existing distrust of the developing countries, and if the US persists in its refusal to ratify the Kyoto Protocol, this is likely to complicate discussions on second commitment period commitments with developing countries. Nevertheless, current political events indicate that the US, in a reversal of its unilateral approach thus far, is seeking global collaboration in its campaign against terrorism. This may imply that it will be difficult for the US to continue to take an isolationist stand in other regime areas, such as climate change policy.

REFERENCES

Agarwal, A. (1999), Personal communication during a visit to RIVM.

Agarwal, A. and S. Narain (1991), *Global Warming in an Unequal World, a Case of Environmental Colonialism*, Delhi, India: Centre for Science and Environment (CSE).

Agarwal, A., S. Narain and A. Sharma (eds) (1999), *Green Politics – Global Environmental Negotiations 1, Centre for Science and Environment*, Delhi, India: Centre for Science and Environment (CSE).

Banuri, T., K. Göran-Mäler, M. Grubb, H.K. Jacobson and F. Yamin (1996), 'Equity and Social Considerations', in J.P. Bruce, H. Lee and E.F. Haites (eds), *Climate Change – 1995 Economic and Social Dimensions of Climate Change*, IPCC, Cambridge, UK: Cambridge University Press.

Baumert, K.A., R. Bhandari and N. Kete (1999), *What Might a Developing Country Commitment Look Like?*, Climate Notes, Washington DC: World

Resources Institute.

Berk, M.M. and M.G.J. den Elzen (1998), 'The Brazilian Proposal Evaluated', *Change*, **44**, 19–23.

Berk, M.M. and M.G.J. den Elzen (2001), 'Options for Differentiation of Future Commitments in Climate Policy: How to Realise Timely Participation to Meet Stringent Climate Targets? *Climate Policy*, **1**(4), 465–80.

Berk, M., J.G. van Minnen, B. Metz and W. Moomaw (2001), *COOL Global Dialogue – Synthesis Report*, Bilthoven, The Netherlands: National Institute for Public Health and the Environment (RIVM) (forthcoming).

Blok, K., G.J.M Phylipsen and J.W. Bode (1997), *The Triptych Approach: Burden Differentiation of CO_2 Emission Reduction Among European Union Member States*, Utrecht, The Netherlands: Utrecht University, Department of Science, Technology and Society.

Centre for Science and Environment (CSE) (1998), *Definitions of Equal Entitlements*, CSE-dossier, fact sheet 5, Delhi, India: CSE.

Claussen, E. and L. McNeilly (1998), *Equity and Global Climate Change, The Complex Elements of Global Fairness*, Arlington, USA: PEW Centre on Global Climate Change.

Den Elzen, M.G.J. and A.P.G. de Moor (2001), *Evaluating the Bonn Agreement and Some Key Issues*, RIVM Report No. 728001/016, Bilthoven, The Netherlands: National Institute for Public Health and the Environment (RIVM).

Den Elzen, M.G.J., M.A. Jansen, J. Rotmans, R.J. Swart and H.J.M. de Vries (1992), 'Allocating Constrained Global Carbon Budgets: Inter-regional and Inter-generational Equity for a Sustainable World', *International Journal of Global Energy Issues*, **2**, 287–301.

Den Elzen, M.G.J., M.M. Berk, M. Schaeffer, O.J. Olivier, C. Hendriks and B. Metz (1999), *The Brazilian Proposal and Other Options for International Burden-Sharing: An Evaluation of Methodological and Policy Aspects Using the FAIR mode*, Report No. 728001011, Bilthoven, The Netherlands: National Institute for Public Health and the Environment (RIVM).

Den Elzen, M.G.J., M. Berk, S. Both, A. Faber and R. Oostenrijk (2001), *FAIR 1.0: An Interactive Model to Explore Options for Differentiation of Future Commitments in International Climate Policy Making*, Report No. 728001012, Bilthoven, The Netherlands: National Institute for Public Health and the Environment (RIVM) (FAIR software can be downloaded via: http://www.rivm.nl/FAIR/.

Depledge, J. (2000), *Tracing the Origins of the Kyoto Protocol: An Article by Article Textual History*, UNFCCC/TP/2000/2, Bonn, Germany: UNFCCC.

Groenenberg, H., D. Phylipsen and K. Blok (2001), 'Differentiating the Burden World-wide – Global Burden Differentiation of GHG Emissions Reductions Based on the Triptych Approach: A Preliminary Assessment', *Energy Policy,* **29**, 1007–30.

Grubb, M. (1989), *The Greenhouse Effect: Negotiating Targets,* London, UK: Royal Institute of International Affairs.

Grubb, M., J. Sibenius, A. Magalhaes and S. Subak (1992), 'Sharing the Burden', in I.M. Mintzer (ed.), *Confronting Climate Change: Risks, Implications and Responses,* Cambridge, UK: Cambridge University Press.

Grübler, A. and N. Nakicenovic (1994), *International Burden-Sharing in Greenhouse Gas Reductions,* Laxenburg, Austria: IIASA.

Gupta, J. (1999), 'North–South Aspects of the Climate Change Issue: Towards a Constructive Negotiating Package for Developing Countries', *Review of European Community and International Environmental Law,* **8**(2), 198–208.

Gupta, J. (2000), *Climate Change: Regime Development and Treaty Implementation in the Context of Unequal Power Relations – Volume 1;* Monograph, Amsterdam, The Netherlands: Institute for Environmental Studies, Vrije Universiteit Amsterdam.

Gupta, J. and N. van der Grijp (1999), 'Leadership in the Climate Change Regime: The European Union in the Looking Glass', *International Journal of Sustainable Development,* **2**(2), 303–22.

Gupta, J., P. van der Werff and F. Gagnon-Lebrun (2001), *Bridging Interest, Classification and Technology Gaps in the Climate Change Regime,* Report E-01/06, Amsterdam, The Netherlands: Institute for Environmental Studies.

Hargrave, T. (1998), *Growth Baselines: Reducing Emissions and Increasing Investment in Developing Countries,* Washington DC, USA: Centre for Clean Air Policy.

IISD (2000), *ENB on the Side,* Special report on selected side events at UNFCCC COP-6 (convened on Saturday 18 November 2000, The Hague, The Netherlands), http://iisd.ca/climate/cop6/side/.

Jacoby, H.D., R. Schmalensee and I. Sue Wing (1999), *Towards a Workable Architecture for Climate Change Negotiations,* Massachusetts Institute of Technology Joint Program on the Science and Policy of Global Change, Report No. 49; Boston, Mass.: MIT Press

Jansen, J.C., J.J. Battjes, J.P.M. Sijm, C. Volkers and J.R. Ybema (2001a), *The Multi-Sector Convergence Approach of Burden-Sharing – A Flexible Sector-based Framework for Negotiating Global Rules for National*

Greenhouse Gas Emission Mitigation Targets, ECN Report, ECN-C--01-007, CICERO Working Paper No. 5, Petten, The Netherlands: ECN/Oslo, Norway: CICERO. (The Multi-Sector Convergence framework software can be downloaded via: http:/www.ecn.nl/unit bs/kyoto/main.html.)

Jansen, J.C., J.J. Battjes, F.T. Ormel, J.P.M. Sijm, C.H. Volkers, J.R. Ybema, A. Torvanger, L. Ringius and A. Underdal (2001b), *Sharing the Burden of Greenhouse Gas Mitigation – Final Report of the Joint CICERO–ECN Project on the Global Differentiation of Emission Mitigation Targets Among Countries*, ECN Report ECN-C--01-009, CICERO Working Paper No. 5, Petten, The Netherlands: ECN/Oslo, Norway: CICERO.

Kawashima, Y. (1996), 'Differentiation of Quantified Emission-Limitation and Reduction Objectives (QUELROs) According to National Circumstances: Introduction to Equality Criteria and Reduction Excess Emissions', Paper presented at 'Informal workshop on quantified limitation and reduction emission objectives', 28 February 1996, Geneva, Switzerland.

Krause, F., W. Bach and J. Koomey (1989), *Energy Policy in the Greenhouse, Vol. 1: From Warming Fate to Warming Limit: Benchmarks for a Global Climate Convention*, El Corrito, USA: International Project for Sustainable Energy Paths.

Meyer, A. (2000), *Contraction and Convergence, the Global Solution to Climate Change*, Bristol, USA: Schumacher Briefings 5, Greenbooks for the Schumacher Society.

Mintzer, I.M. and J.A. Leonard (eds) (1995*), Negotiating Climate Change: The Inside Story of the Rio Convention*, Cambridge, UK: Cambridge University Press.

Müller, B. (1999), *Justice in Global Warming Negotiations – How to Achieve a Procedurally Fair Compromise*, Oxford, UK: Oxford Institute for Energy Studies.

Müller, B., A. Michaelova and C. Vrolijk (2001), *Rejecting Kyoto, A Study of Proposed Alternatives to the Kyoto Protocol* (Pre-publication copy for general distribution), Hamburg/Oxford/London: Climate Strategies International Network for Climate Policy Analysis.

Mwandosya, M.J. (1999), *Survival Emissions: A Perspective from the South on Global Climate Change Negotiation*, Tanzania: DUP (1996) LIMITED, and The Centre for Energy, Environment, Science and Technology (CEEST-2000).

Philibert, C. (2000), 'How Could Emissions Trading Benefit Developing Countries', *Energy Policy*, **28**, 947–56.

Philibert, C. and J. Pershing (2001), 'Considering the Options: Climate Targets for All Countries', *Climate Policy*, **2**, 211–27.

Phylipsen, G.J.M., J.W. Bode, K. Blok, H. Merkus and B. Metz (1998), 'A Triptych Sectoral Approach to Burden-sharing; Greenhouse Gas Emissions in the European Bubble', *Energy Policy*, **26**, 929–43.

Ringius, L., A. Torvanger and B. Holtsmark (1998), *Can Multi-criteria Rules Fairly Distribute Climate Burdens? OECD Results from Three Burden-sharing Rules*, CICERO Working Paper 1998: 6, Oslo, Norway: CICERO.

Ringius, L., A. Torvanger and A. Underdal, A. (1999), *Burden Differentiation of Greenhouse Gas Abatement: Fairness Principles and Proposals*, The joint CICERO–ECN project on sharing the burden of greenhouse gas reduction, among countries, Report ECN-C--00-011/ CICERO WP 1999: 13, Petten, The Netherlands: ECN/Oslo, Norway: CICERO.

Rose, A. (1992), 'Equity Considerations of Tradable Carbon Emission Entitlements', in *Combating Global Warming*, Geneva, Switzerland: UNCTAD.

Rose, A., B. Stevens, J. Edmonds and M. Wise (1998), 'International Equity and Differentiation in Global Warming Policy', *Environmental and Resource Economics*, **5**, 193–208.

Sijm, J., J. Jansen and A. Torvanger (2001), 'Differentiation of Mitigation Commitments: The Multi-sector Convergence Approach', *Climate Policy*, **1**(4), 481–97.

Torvanger, A. and O. Godal (1999), 'A Survey of Differentiation Methods for National Greenhouse Gas Reduction Targets', CICERO Report 1999: 5, Oslo, Norway: University of Oslo.

Torvanger, A. and L. Ringius (2000), *Burden Differentiation: Criteria for Evaluation and Development of Burden-sharing Rules*, The joint CICERO–ECN project on sharing the burden of greenhouse gas reduction among countries, Report ECN-C--00-013/CICERO WP 2000: 1, Petten, The Netherlands: ECN/Oslo, Norway: CICERO.

Torvanger, A., T. Berntsen, J.S. Fuglestvedt, B. Holtsmark, L. Ringius and A. Aaheim (1996), *Exploring Distribution of Commitments – A Follow-up to the Berlin Mandate*, CICERO Report 1996: 3, Olso, Norway: CICERO.

UNFCCC (1995), *Conference of the Parties, First Session*, Berlin, Germany, 28 March – 7 April 1995, FCCC/CP/1995/7, Bonn, Germany: UNFCCC/ UNEP.

UNFCCC (1997), *Paper No. 1: Brazil; Proposed Elements of a Protocol to the United Nations Framework Convention on Climate Change, Presented by Brazil in Response to the Berlin Mandate*, UNFCCC/AGBM/ 1997/MISC.1/Add.3 GE.97-, Bonn, Germany: UNFCCC.

UNFCCC (1998), *The Kyoto Protocol to the Convention on Climate Change*, UNEP/IUC/98/2, Châtelaine, Switzerland: UNFCCC/ UNEP/IUC.

UNFCCC (2001a), *Implementation of the Buenos Aires Plan of Action*, FCCC/CP/2001/L.7, 24 July, Bonn, Germany: UNFCCC.

UNFCCC (2001b), 'The Marrakesh Accords and the Marrakesh Declaration' (advance version of decisions and other actions adopted by the Conference of Parties to the UNFCCC on its seventh session (COP-7)), Bonn, Germany: UNFCCC (published at web site only: http:/www/unfccc/int/).

10. Linkages between the climate change regime and the international trade regime

Onno Kuik, Richard S.J. Tol and David J.-E. Grimeaud[1]

1 INTRODUCTION

The international climate change regime has many linkages to other international regimes, notably other international environmental agreements and the world trade regime as governed by the World Trade Organization (WTO). The linkages can point to (potential) synergies between regimes, but also to contradictions.

This chapter aims to highlight certain linkages to the international trade regime. For linkages to other regimes, we refer to Sands (2000), who gives a comprehensive overview of all linkages between international environmental issues. Section 2 discusses legal linkages to the world trade regime. Section 3 discusses material linkages to the trade regime as they may manifest themselves in changes in international competitiveness and patterns of trade, and Section 4 offers conclusions.

2 THE WORLD TRADE REGIME

The linkage between the international climate change regime and international trade law (GATT/WTO law) will certainly prove to be one of the most sensitive implementing concerns among parties and non-parties alike.

In fact, more than 20 multilateral environmental agreements (MEAs) have been identified as containing explicit trade-restrictive measures. They include the 1973 Convention on International Trade in Endangered Species of Wild Fauna and Flora (CITES), the 1987 Montreal Protocol and the 1989 Basel Convention on Transboundary Movements of Hazardous Wastes and Their

Disposal. In that regard, one should note that no clear-cut conclusions have so far been reached on how conflicts that may arise between GATT/WTO law and environmental measures taken pursuant to a MEA should be resolved. Nor has a conclusion been reached on the appropriate settlement mechanism or forum in the event of a dispute, in particular when the two opposing parties are both members of the WTO but where only one of them is party to the relevant MEA (WTO, 1996). In practice, however, no such conflicts have yet occurred. Several reasons might be identified, including the fact that the above-mentioned international environmental agreements result from a wide consensus on the need to urgently address issues such as the protection of endangered species, ozone depletion and the disposal in developing countries of hazardous waste. In addition, it may be argued that compliance with the relevant MEAs has not had a major economic or social impact, while environmental benefits are clearly identifiable.

However, many economists claim that the entry into force and implementation of the international climate change regime, including provisions of the UNFCCC and in particular of the Kyoto Protocol will undoubtedly bring about great economic and social changes. Understandably, compliance with Article 3(1) of the Kyoto Protocol on the reduction of CO_2 emissions will call for fundamental alterations in, for instance, national energy, industrial and transport policies. In this context, the long-standing opposition by petroleum-exporting states and the refusal of the US administration to ratify the Protocol show how economically and socially sensitive the global warming issue is.

Yet, by contrast with the above-mentioned MEAs, the international climate regime does not in itself contain explicit trade-restrictive measures. As we discuss below, Article 17 of the Kyoto Protocol on emissions trading simply lays down general rules, while parties remain responsible for designing emissions trading schemes and organizing the participation of legal entities. As regards policies and measures to be adopted pursuant to Article 2. parties are allowed a great deal of freedom to decide what they should do domestically to reduce greenhouse gas emissions. As far as the potential effect they will have on trade is concerned, Article 3(5) of the UNFCCC 'Principles' provides that measures that are taken to combat climate change, including unilateral ones, should not constitute a means of arbitrary or unjustifiable discrimination or a disguised restriction on international trade. Similarly, Article 2(3) provides that Annex I parties shall 'strive to implement policies and measures in such a way as to minimize ... effects on international trade'. The climate change regime is thus clearly intended to develop within the international trade law framework, whereby environmental and trade concerns may both be addressed, but within the scope of the rules and principles that govern trade liberalization, including in particular Articles I, III, XI and XX of GATT.

Before examining the types of international trade law conflicts that may arise from the implementation of the international climate change regime, and of the limits within which domestic measures may be adopted, it is first necessary to look at the relevant GATT/WTO law principles and rules. We also refer to some of the key GATT panels in order to shed light on how they have been interpreted. We then provide an overview of the particular trade problems that might arise from national climate change policies, from the practical implementation of emissions trading schemes, and from the enacting of policies and measures under Article 2 of the Kyoto Protocol. For a detailed analysis of the world trade regime and the environmental issue, see Calster (2000).

2.1 GATT/WTO Law and Principles that may be Relevant to Environmental Measures

When implementing the international climate change regime and designing related national policies, Annex I parties have to consider provisions of the GATT, in particular Articles I, III, XI and XX. In addition, if a national measure qualifies as a 'technical regulation', Article 2 of the Agreement on Technical Barriers to Trade (TBT) may apply.

Articles I, III and XI GATT
With regard to provisions that fall under the GATT, Articles I and III prescribe 'relative standards' that are non-discriminatory. More specifically, Article I GATT (Most Favoured Nation principle) states that WTO members shall impose the same fiscal or regulatory treatment on 'like' products, whatever their origin. No discrimination shall take place among imports and exports on the basis of the country of origin or of the country of destination. In other words, the granting of MFN treatment to one WTO member shall benefit all other WTO members.[2] Thus, where an Annex I Party sets, for example, tax-based or regulatory standards for energy products, the same treatment shall benefit all 'like' imports.

In addition to the MFN principle, Article III GATT provides that any fiscal or regulatory treatment that is imposed on imports shall be imposed equally on 'like' domestic products (National Treatment principle). More specifically, Article III(2) GATT holds that imports shall not be subject to taxes or charges in excess of those imposed on 'like' or 'directly competitive or substitutable' national products. As far as regulatory provisions are concerned, Article III(4) provides that imports shall not be accorded less favourable treatment than that accorded to 'like' products of national origin. This relates to the national measures that set requirements affecting the internal sale, offering for sale, purchase, transportation, distribution and use of products. Annex I parties shall therefore not be entitled to discriminate under Article III (2) and (4) against imports *vis-à-vis* 'like' domestic products when enacting national climate change policies and measures. In addition, internal taxes and regulations should not be applied in a way that would afford protection to domestic production (Article III(1) GATT). In other words, beyond mere equality of fiscal or regulatory treatment, they may not be applied with the objective of benefiting national products. The distortion of competition between imports and domestic goods will be held to be incompatible with GATT if similar tax-based or regulatory measures are adopted precisely and purposely for the protection of national production.

One central issue in the application of Articles I and III GATT relates to the concept of the 'likeness' of products. Indeed, where two products are considered to be 'like' products, neither may be subject to differentiated treatment. In practice, GATT panels seem to have used a set of four main criteria on a case-by-case basis. These criteria are tariff classification by GATT members, the product's end-uses on a given market, the consumers' tastes and habits and the products' properties, nature and qualities.[3] In this regard one should recall that production processes and methods (non-incorporated PPMs) that do not alter the physical properties of a given product are not taken into consideration when one examines 'likeness'. Thus, an Annex I Party shall not be entitled to adopt a different fiscal or regulatory treatment, for instance for imported and domestically produced electricity, on the basis of its mode of generation. No consideration shall be given to the fact that it might have been produced by using renewable energy sources, or conversely by fossil-fuel burning units.

As an 'absolute' standard, Article XI GATT provides that WTO members must not establish import or export prohibitions or restrictions via measures other than taxes or charges, such as quantitative quotas or licences. In this context, it remains crucial to distinguish those national measures that would fall under either Article XI or Article III GATT. Indeed, a trade-restrictive regulatory national environmental measure that would fall under Article III GATT may be justifiable provided that it complies with Article I(1) GATT on MFN, Article III(1) GATT on the avoidance of protection for national production and Article III(4) GATT on similar treatment for 'like' products. However, if Article XI GATT applies, any restriction on imports or exports is prohibited, which will require immediate recourse to Article XX GATT.

Article XX GATT

Whether or not provisions of a national climate change policy are to be considered in breach of Articles I, III or XI GATT, they may still be justifiable under Article XX (b) or (g) GATT, provided that certain conditions are met.[4]

To rely on Article XX(b) GATT, the party 'invoking' it would have to demonstrate that its policy in respect of which the measures were adopted fell within the range of those that are designed to protect human, animal or plant life or health. In this regard, one may recall that in the 1996 US *Gasoline* case, the panel ruled that a policy to reduce air pollution resulting from the consumption of gasoline was a policy falling within the range of the ones concerning protection of human, animal and plant life or health, as well as within the scope of a policy aimed at conserving a natural resource (air) within the meaning of Article XX(g) GATT.[5] Therefore, it might be argued that a national climate change policy that sought to reduce greenhouse gas emissions could be considered in the light of Article XX(b) and/or (g) GATT. Second, the WTO member concerned would have to demonstrate that its measure is 'necessary' to achieve the relevant policy objective. Although the policy objective in itself (its scope) is not subject to scrutiny, the measures employed to realize it must be those that are reasonably available and the least-restrictive.[6]

In the event of recourse to Article XX(g) GATT, the 'defending' WTO member would have to show that the relevant measure 'relates to the conservation of exhaustible natural resources'. The term 'relates' has been interpreted either as 'primarily aimed at'[7] or as 'a close and genuine relationship of ends and means'[8] and taken to mean that the measure is put into effect 'in conjunction with restrictions on domestic production or consumption', which implies that relevant domestic products or activities are also subject to restrictions.[9]

In addition, the 'defending' WTO member would have to show that its measures are consistent with the conditions of the chapeau or Headnote of Article XX. In particular, the environmental measures would have to be applied in a manner that does not constitute an unjustifiable or arbitrary discrimination between countries where similar conditions prevail, or a disguised restriction on international trade. In fact, although the terms 'arbitrary discrimination', 'unjustifiable discrimination' and 'disguised restriction' have not been precisely defined, their purpose is to avoid abuses or illegitimate use of the exceptions to the substantive rules set out in Article XX GATT.[10]

Article 2 TBT

Besides the above-mentioned GATT principles, a national measure may also fall under the TBT if, for example, it qualifies as a 'technical regulation'.[11] In such a case, under Article 2(1), the WTO member would have to ensure that it accords treatment no less favourable than that accorded to 'like' national products (National Treatment principle) and to 'like' products originating in any other country (Most Favoured Nation principle). In addition, whereas such regulations may be adopted 'to protect human health or safety, animal or plant life or health, or the environment', regulations shall not be applied in a manner that creates unnecessary obstacles to international trade, and they should also be the least trade-restrictive ones (Article 2(2)).

Therefore, from a general perspective, an Annex I Party may be entitled to enact national climate change policies and measures that might result in trade restrictions provided that it complies with Article I, III or XI GATT. In particular, they must not be discriminatory or protective. However, even if one of those provisions is breached, they may still be 'saved' by virtue of Article XX(b) or (g) GATT, unless the conditions set out in those sections are not complied with. In that respect, one of the key issues for the WTO member concerned would be to use the least trade-restrictive environmental measures available which would permit it to achieve its policy objective. Similarly, trade-restrictive national environmental measures may be considered compatible with the TBT provided that the conditions set out in Article 2 TBT are met.

Within this overall context, a central question remains regarding the ability of WTO members to enact national measures aimed at protecting human, animal or plant life or health (under Article XX(b) GATT), or at conserving an exhaustible natural resource (under Article XX(g) GATT) located outside their national jurisdiction. In other words, one may wonder how a WTO panel might assess a trade-restrictive climate change measure that would also apply extraterritorially. In fact, it might be argued that the GATT jurisprudence may have somewhat slightly evolved towards a more environmentally friendly perspective following the 1996 *Shrimp-Turtle* case. Specifically, the Appellate Body held that 'it is not necessary to assume that requiring from exporting countries compliance with, or adoption of, certain policies ... prescribed by the importing country, renders a measure incapable of justification under Article XX'.[12] Thus, contrary to the findings in the previous *Tuna-Dolphin* cases I and II – where it had been ruled that a national measure which would force another country to alter its production policy would have to be regarded as incompatible with GATT – it might be held that a conservation policy such as that adopted by the US administration to protect sea-turtles, which required the use of 'turtle excluder-devices', might not be automatically held GATT-incompatible, because the aim was to protect a common exhaustible natural resource found in both territorial and extraterritorial waters.[13] In fact, the US policy failed to meet the requirement of the Headnote of Article XX GATT on the grounds that the way it had been implemented amounted to unjustifiable and arbitrary discrimination. In this context, the Appellate Body recalled that an essential element for GATT justification relates to the need to try to reach international or regional agreements with potentially affected parties before any trade-restrictive national environmental measures are enacted.[14] If no such attempts are made, panels will certainly consider the national measure to be 'unilateralism', posing unjustifiable trade restrictions. With regard to the PPMs issue ('turtle excluder-devices' as non-incorporated PPMs), one should note that the Appellate Body's conclusions on Article XX GATT do not address them, in contrast to the panel report that made a clear distinction between product standards and PPMs. Nevertheless, one should wait for future cases as it is still too early to draw conclusions on non-incorporated PPMs issues, at least with regard to protection of shared natural resources, including the air.

2.2 Potential Linkages Between GATT/WTO Law and Climate Change Policies

This section provides a brief, non-exhaustive overview of the potential conflicts that may arise between national climate change policies adopted pursuant to the UNFCCC and the Kyoto Protocol and the international trade law principles and rules discussed above.[15]

Climate-related policies and measures

Article 2 of the Kyoto Protocol sets out a non-compulsory list of policies and measures that should be undertaken by Annex I parties to reduce greenhouse gas emissions domestically. Within this framework, one may look at several instruments, including environmental taxes or charges, product energy efficiency and emission standards and environmental subsidies.

For example, regarding carbon taxes, one might assume that an Annex I party would want to impose a charge or border tax adjustment on imported energy-related products such as fuel, with a view to reducing its overall consumption. According to Article III(2) GATT (see above), it will be entitled to do so provided it is not discriminatory (it will apply in the same manner to all 'like' domestic and imported energy-related products) and that it is not set in such a way as to protect national production.[16] In this context, one should also recall that Article II, (2)(a) GATT provides that a tax may be imposed on an article from which the imported product has been manufactured or produced in whole or in part. This implies that an Annex I party would be able to tax goods on the basis of the volume of fuel used to manufacture it.[17] However, one should recall that up to now non-incorporated PPMs have not been considered as a factor when considering 'likeness'. In short, an Annex I party is not entitled to impose a charge on an imported product based on the amount of fuel used in its production process.

Along with border tax adjustments, one might expect Annex I parties to impose emissions and energy-efficiency standards on products sold in its domestic market. The first concern would be to determine whether two products with different emission or energy efficiency patterns would be considered to be 'like' products.[18] If they were, the regulating country would not be entitled to impose a less favourable treatment on the most polluting items, either under Article III(4) GATT or under Article 2(1) TBT. In this case, the country would have to have recourse to Article XX GATT or to Article 2(2) TBT to 'save' the measure.

An Annex I party might also think of providing subsidies to certain sectors or plants to promote the use of cleaner technology, or to support the production and sale of the most climate-friendly products. Should this be the case, it would then have to ensure compliance with the provisions of the 1994 Agreement on Subsidies and Countervailing Measures (SCM). In particular, national subsidies must not fall into the 'prohibited subsidies' category, namely subsidies that seek to promote the use of domestic goods to the detriment of imports (Article 3(1)(b)). Otherwise, it may result in counter-measures being imposed on that country by the affected state. In addition, they must not qualify as 'actionable subsidies', which include any subsidy that would adversely affect the interests of another WTO member (Article 5). In fact, 'non-actionable subsidies' may be allowed under Article 8(1). These are subsidies that are designed to help promote, under certain conditions, the adaptation of existing facilities to new environmental requirements imposed by law which result in greater constraints and financial burdens (Article 8(2)(c)).

Emissions trading schemes

Article 17 of the Kyoto Protocol allows Annex I parties to achieve parts of their Assigned Amounts (AAs) through the use of an emissions trading scheme. In this context, Articles 3(10) and (11) of the Kyoto Protocol stipulate that such trading shall result in Emission Reduction Units (ERUs) or Parts of Assigned Amounts (PAAs) being added to the AAs of the party who buys them, and conversely being subtracted from the AAs of the party who sells them. In this regard, Annex I parties may have to be cautious, when designing a national emissions trading scheme and when organizing their participation in a wider international scheme, not to violate the GATT/WTO law and principles discussed above.[19]

In fact, most authors would argue that an allowance like a government-issued permit is not a product under GATT, which only covers 'tangible goods' such as raw materials (including energy and energy-related products like oil, gas, electricity) and manufactured goods. Thus, Annex I parties would be free to limit and organize any trade in allowances between regulated national companies and foreign undertakings. An Annex I party would be entitled to prohibit such trade occurring with companies located in a country that does not fulfil its commitments under the Kyoto Protocol. It would also be allowed to restrict the amount of emission allowances that would be allowed domestically.[20]

On the other hand, if these national rules on emission trading are not carefully designed, they might negatively affect the international trade in energy and energy-related products, which should in fact benefit from the application of GATT/WTO rules and principles. In fact, given that the amount of national emission allowances would certainly be limited for each Annex I party, the importers or the sellers of energy-related products (who would be required to hold such permits in order to be allowed to bring their products onto national markets) may have to limit the volume of their imports. This might in turn be interpreted as a breach of Article XI GATT. Accordingly, national authorities would have to ensure that foreign importers or sellers enjoy a non-discriminatory allocation of national emission allowances, either through grandfathering or auctioning, so as to provide equal opportunities for imported products *vis-à-vis* national products. This would be all the more important for newcomers, who should not be deprived of the ability to enter national energy-related product markets. Should this not be the case, there would then be a risk of violating either Article I or Article III GATT, as well as Article XI GATT. A related issue concerns the criteria upon which emission allowances would be allocated. As discussed above, an Annex I party is not entitled to allocate allowances on the basis of the mode of generation of electricity, for example, as that would amount to taking into consideration non-incorporated PPMs. On the other hand, one may wonder to what extent such a country might regulate the allocation of emission permits based on the volume of greenhouse gas emissions that would result from the consumption of fuel, natural gas or petroleum products. One of the key concerns would be to determine whether those energy-related products would be seen as 'like' products and whether this would intentionally benefit national production. It might also violate Article 3 or Article 5 of the 1994 SCM.

3   PATTERNS OF INTERNATIONAL TRADE

The international climate change regime has linkages with economic activities and with patterns of international trade. Reduction of emissions of greenhouse gases in one group of countries, as agreed upon in the Kyoto Protocol, might affect worldwide patterns of production and consumption of energy and non-energy commodities, and the pattern of international trade.

3.1   Economic and Trade Impacts of CO_2 Reduction Policies

In the economic literature, attention has primarily focused on the interaction between CO_2 reduction policies and the energy market.[21] CO_2 reduction policies in Annex B countries directly or indirectly increase the producers' price of energy and energy services, thereby resulting in adjustment costs. These adjustment costs depend not only on the level and speed of the necessary adjustments, but also on the flexibility of the economy, the availability and costs of alternatives, the efficiency of the policy employed, and a number of other characteristics of the economy. There are considerable uncertainties surrounding many of the above factors and it is not surprising that estimates of the economic impact of CO_2 reduction policies vary considerably.

Let us suppose that countries that want to reduce their CO_2 emissions (Annex B countries) levy a tax on these emissions. Firms and households are confronted with a tax on fuels that increases with the CO_2-intensity of the fuels. They have an incentive to switch to less CO_2-intensive fuels, or to use more energy-efficient technologies. The demand for CO_2-intensive fuels in Annex B countries falls. Unless there are energy-saving options with negative or zero costs, firms and households see their fuel bills (inclusive of taxes) rise. The extent of this rise depends on the rate of the tax, the impact of the tax on the market price of fuels, and their ability to switch to less carbon-intensive fuels and apply energy-saving technologies. This effect on the costs of the use of fuel has effects in the markets of non-energy goods and services as well.

The cost of producing goods and services that are relatively CO_2-intensive in their production – such as iron and steel, non-ferrous metals, and chemicals – will increase relative to the production costs of goods and services that are less carbon-intensive. Output of CO_2-intensive industries will fall, while output of less CO_2-intensive industries (for example agriculture, services) will rise as labour and capital services are released from CO_2-intensive industries and find alternative employment in less CO_2-

intensive industries. Despite this compensating growth in less CO_2-intensive industries, Gross Domestic Product (GDP) will probably fall.

The economic impact of CO_2 reduction policies in Annex B countries on non-Annex B or developing countries is transmitted through the international markets for energy and non-energy commodities, and the international markets of factors of production.

As noted above, Annex B countries' demand for carbon-intensive fuels falls. This directly affects the exports of fossil fuel-producing developing countries. Depending upon the elasticity of the supply of fossil fuel producers,[22] the world price of CO_2-intensive fossil fuels may drop. When not confronted with CO_2 taxes, firms and households in developing countries see the prices of CO_2-intensive fuels fall and may find it profitable to increase the use of these fuels. There is less incentive to save energy. This additional demand for CO_2-intensive fuels offsets to a certain extent the drop in world market prices of fuels.

As in the Annex B countries, in developing countries there may also be secondary impacts on the markets of non-energy goods and services. These impacts mirror the impacts in Annex B countries discussed above. The costs of producing goods and services that are relatively CO_2-intensive in their production will fall relative to the production costs of goods and services that are less carbon-intensive. Moreover, the costs of CO_2-intensive goods and services in developing countries will fall relative to the costs of their competitors in Annex B countries.

In the longer term, domestic and global investments may adjust to the changed set of prices on the markets in Annex B and non-Annex B countries. This may cause a relative shift of investments in CO_2-intensive industries from countries with cost-increasing CO_2 reduction policies towards countries without such policies. This would lead to an expansion of CO_2-intensive production in developing countries. Finally, in the long term the dynamics of technological innovation and development may have an effect on the growth of developing countries' economies. These dynamics, discussed below, could lead to an alleviation of current energy supply constraints.

3.2 Terms of Trade, Leakage and Technology Dynamics

The overall economic and environmental consequences of these effects can be positive, neutral or negative, depending on the manifestations and relative scale of the effects. This is an empirical matter to which we turn later, but first we shall explain the terms-of-trade effect, the carbon-leakage effect, the growth effect and the technology-dynamics effect in more detail.

Terms of trade

A country's 'terms of trade' are defined as the ratio of aggregate export prices to aggregate import prices. If a country's terms of trade fall, it can get less import goods in exchange for its export goods. It was noted above that the price of fossil fuels tends to fall on the world market. A price reduction of other commodities that are intensively used in CO_2-intensive industries (metals, minerals) can also be expected. On the one hand, this will negatively affect the terms of trade of countries that rely heavily on the exports of these commodities, especially oil-exporting developing countries. On the other hand, a reduction in the price of fossil fuels may improve the terms of trade of developing countries that import fossil fuels. Hence, there is an important difference between developing countries that export fossil fuels and developing countries that import fossil fuels. In the non-energy goods markets, it is likely that the import prices of CO_2-intensive goods will go up because of the market power of industries in Annex B countries. Most notably, the price of chemicals, iron and steel, rubber and plastics, and non-ferrous metals are likely to rise.

Carbon leakage

The term 'carbon leakage' is used to describe the effect when some countries abate CO_2 emissions while CO_2 emissions increase in non-abating countries as a consequence of changes in world market prices of energy commodities, or as a consequence of the relocation of energy-intensive industries to these countries. The mechanism underlying this effect was discussed in Section 3.1. The increase of emissions in non-Annex B countries as a fraction of the reduction of emissions in Annex B countries is called carbon leakage.

Based on an extensive review of literature, the IPCC (2001) recently reported a credible range between 5 and 20 per cent for carbon leakage in its Third Assessment Report. It noted that, all other things being equal, the rate of carbon leakage might be less if:[23]

- more use is made of flexible instruments that reduce abatement costs;
- OPEC is effective in adjusting the supply of oil in order to keep its price up; and
- there is more transfer of environmentally sound technologies and know-how to developing countries.

Model results on carbon leakage need to be interpreted with care, as international capital markets occasionally defy economic theory and companies' location decisions are complex and context-dependent.

Economic growth and imports

As argued above, greenhouse gas emission reduction may reduce economic growth in Annex B countries. If growth slows, so does consumption and imports. This slowdown in world trade would hurt exporting developing countries. IPCC (2001) reports that the majority of studies show reductions of projected GDP of about 0.2 to 2 per cent in 2010 for OECD countries without emissions trading, and about 0.1 to 1.1 per cent of projected GDP with emissions trading. The effect of the Kyoto Protocol on the growth of exports of developing countries is therefore likely to be limited.

Technology dynamics

In the longer term, reducing carbon dioxide emissions would change the technological and commercial structure of energy supply and demand. It may be expected that, because of the higher costs of using fossil fuels, energy and carbon-saving technologies would increase in performance and decrease in price through learning by doing and research and development (R&D). If, however, a significant number of companies decide to relocate their production to countries with less stringent climate policies (see above), the stimulus for technological development would be severely reduced, since it is reasonable to assume that the companies that suffer most from emission abatement are the first to move, if that is an option, or contribute most to R&D, if moving is not an option.

More rapid technological progress would have three effects. First, the costs of emission reduction in the Annex B countries would go down, thereby weakening its effects on international capital allocation and economic growth. Second, the costs of any future emission reduction in developing countries would decline provided that the newly developed techniques are available on the world market, which is likely, and provided that the new technology is affordable to poorer countries, which is uncertain. Third, the range of energy-supply technologies (including energy saving) would increase both technologically and economically. For many developing economies energy is currently a constraint, for example, electricity generating capacity or fossil fuel imports. Through technological development and diffusion, emission reduction policies in Annex B countries would help alleviate such constraints. The impact on emissions depends on the relative scale of increases in the demand for energy and improved energy efficiency and in the share of renewable energy.

The literature is starting to come to grips with greenhouse gas emission reduction and induced technological change. The complex interplay between international investment, leakage and technological development is barely understood (cf. Tol et al., 2001, for some first steps).

Besides international investments, international trade also affects which technologies are used and developed. Differences in domestic regulation (for example, energy efficiency standards) could lead to fragmentation of the international market. However, scale economies may lead to overall adoption of more stringent targets. This is observed with cars, for instance. The tight regulations in Japan (on energy efficiency) and California (on a range of pollutants), both substantial car markets, also improve the environmental performance of cars in more loosely-regulated markets, for instance in Europe. This effect is unambiguously positive (for the environment).

There is also international trade in second-hand goods. Tighter regulation would likely push up exports of older, dirtier goods. In the importing countries, these would partly replace even older, dirtier goods. It would also increase supply and depress prices, which would increase overall consumption. In the short term, therefore, the environmental effect may be positive as well as negative. In the long term, the environmental effect is positive since importing countries automatically adopt the standards of exporting countries, with the lag determined by the technical lifetime of the product in question. (Travelling through Africa or Asia, casual observation shows that the technical lifetime of cars, for example, can be remarkably long.)

At the moment, the interactions between international trade and technology diffusion are known in theory, but not in practice. Empirical and modelling studies to estimate the strength of the mechanisms sketched above are lacking.

3.3   Emissions Trading, Joint Implementation and the Clean Development Mechanism

The Kyoto Protocol allows countries to co-operate in fulfilling their commitments. In particular, the Protocol distinguishes Emissions Trading, Joint Implementation and the Clean Development Mechanism. Their exact nature is discussed in Chapter 7 above by Woerdman et al.

The potential effects of these 'flexibility mechanisms' on international trade are threefold. In the first place, these mechanisms may reduce the compliance costs of Annex B countries in meeting their commitments. In fact, that is what they were intended for. If compliance costs are lower, there will be fewer adjustments necessary in energy and non-energy markets. Hence, there will be less of an impact on relative prices, trade flows and economic welfare.

Second, flexibility mechanisms imply capital transfers. The generated consumer and producer surpluses, and collected monopoly and monopsony rents will affect countries' incomes, and thus consumption and the demand for imports.

Finally, emissions trading can have an impact on the terms of trade of the trading partners. OECD (1999) lists three reasons why permit buyers could experience adverse terms-of-trade effects. First, in order to buy permits from abroad, a country has to increase its exports of goods and services (or reduce its imports) if it does not want to run a deficit on its balance of payments. It can do so by lowering its real exchange rate, reducing the price of exports and increasing the price of imports. Second, buying permits from abroad allows a country to use, and consequently import, more oil than without the option of emissions trading. To be able to import more oil, a country has to increase its exports or reduce its imports of other goods and services, leading to the same adjustment process as above. Finally, emissions trading among Annex B countries shifts abatement from coal to oil, increasing the international price of oil. In a theoretical paper, Copeland and Taylor (2000) show that, under certain conditions, a country can even be worse off with emissions trading if that trading causes a sufficiently large deterioration of its terms of trade. This is an application of the well known 'immiserizing growth' theorem of Bhagwati (1958). In empirical assessment, however, the adverse terms-of-trade effect reduces, but does not dissipate, the gains from emissions trading.

3.4   Assessment of Overall Trade Impacts

Indications of the overall effect of carbon-reduction policies on the economies of Annex B and developing countries and on global CO_2 emissions are provided by a number of economic model computations that were presented at the 16th Energy Modelling Forum at Stanford University (Weyant, 1999) and, more recently, by IPCC's Third Assessment Report (IPCC, 2001). The models differ greatly in terms of structure and assumptions about exogenous developments, such as the growth in CO_2 emissions in the absence of restrictive measures. As indicated above, there is also a great deal of uncertainty about the costs of reducing emissions, in particular those of non-CO_2 greenhouse gases.[24]

Impacts on the competitiveness of specific industries
As mentioned in Section 3.2, the overall impact on the GDP of Annex B countries is generally assumed to be limited (IPCC, 2001). This does not exclude potentially larger effects on specific energy and trade-intensive industries. Some models, for example the MERGE model of Manne and Richels (1999), predict a relatively severe impact on the competitiveness of

energy-intensive industries in Annex B countries, in the order of magnitude of a 10 per cent fall in output. Others, though, for example GREEN (OECD, 1999), forecast only relatively small effects (–1 per cent in EU and Japan, –3 per cent in USA and –4 per cent in the rest of the OECD). The MS-MRT model of Bernstein et al. (1999) projects a relatively large impact on the energy-intensive industries of the USA (–7.9 per cent), moderate effects on other OECD (–2.7 per cent), but minor effects on Japan (–1 per cent) and the EU (–0.2 per cent). IPCC (2001) concludes from a thorough review of literature that the estimated effects on price competitiveness are small.

Energy-intensive industries in economies in transition may gain a competitive advantage over their competitors in the rest of the Annex B region. This is because the Kyoto targets for economies in transition (especially the former Soviet Union) are generally expected to be mild or even non-binding, and because their marginal abatement costs are generally considered to be very low.

The overall, trade-induced impact of Kyoto on the economies of developing countries is mixed, see Table 10.1. The impact on their energy-intensive industries is generally positive. But because of the negative impact of Kyoto on fuel prices, energy-exporting countries may see their terms of trade worsen to such an extent that they suffer on balance. This would be particularly true for the Middle East, and for a country like Mexico. Net energy importers, on the other hand, profit from the combination of lower fuel prices and the increased competitiveness of their energy-intensive industries. This would be true for countries such as Korea, India and Brazil.[25] Model assessments of these effects generally forecast small overall changes in either direction, while they are indecisive for certain countries such as China, where different models predict different signs of change. Emissions trading among Annex B countries would generally reduce both the positive and negative impacts on GDP of developing countries. In some cases, no effects of emissions trading can be discerned.

For most countries, the impact on GDP seems to be very limited. To put that into perspective, a 1 per cent reduction of GDP in 2010 is identical to a slowdown in the growth rate of 0.06 per cent over the period 2000–2010. If average growth without Kyoto would have been 5 per cent per year, the trade-induced impact of Kyoto would slow this growth to 4.94 per cent per year. Most studies forecast the impact on GDP to be less than 1 per cent for most countries in either direction.

Table 10.1   *Effects (%) of Kyoto Protocol on GDP of some developing countries in 2010, with no emissions trading and with emissions trading among Annex B countries (between brackets)*

	AIM	MS-MRT	GTEM	G-Cubed	WorldScan	GREEN
China	–0.2 (–0.1)	0.3 (0.1)	–0.2 (–0.1)	–0.2 (–0.1)		
India	+0.3 (+0.1)	0.6 (0.5)[1]	0.0 (0.0)			
Korea	+0.6 (+0.2)	0.5 (0.5)[2]	0.0 (0.0)		–0.1 (–0.1)[2]	
Middle East	–1.5 (–0.6)	–1.0 (–0.6)[3]			–1.0 (–0.6)	
Mexico		–1.0 (–0.6)[3]	–0.2 (0.0)		–0.3 (–0.2)[4]	
Brazil			0.1 (0.1)			
Non-Annex B			0.0 (0.0)	0.4 (0.4)[5]	–0.3 (–0.2)	–0.2 (–0.1)

Notes:
1. India and China.
2. South-East Asia (incl. Korea).
3. Mexico and OPEC.
4. Latin America.
5. Excluding China.
AIM: Asian-Pacific Integrated Model, Kainuma et al. (1999).
MS-MRT: Multi-Sector Multi-Region Trade Model, Bernstein et al. (1999).
GTEM: Global Trade and Environment Model, Tulpulé et al. (1999).
G-Cubed: Global General Equilibrium Growth Model, McKibbin et al. (1999).
WorldScan: Dynamic Model for the World Economy, Bollen et al. (2000).
GREEN: General Equilibrium Environmental Model, OECD (1999).

4   CONCLUSIONS

The international climate change regime has many linkages to other international regimes, most notably other international environmental agreements and the world trade regime.

A widely discussed linkage is that between the climate change regime and the world trade regime. Linkages can be found at the level of principles, norms, rules and decision-making procedures, as well as at the material (economic) level. The major principle of the world trade regime is non-discrimination between different foreign suppliers and between foreign and domestic suppliers of a like product. The domestic implementation of policies to reduce greenhouse gases in accordance with the obligations of the Kyoto Protocol could lead to conflicts with the world trade regime if such policies were applied in such a manner that they would discriminate between different foreign suppliers or between domestic and foreign suppliers. Whether this is a positive or a negative linkage from the point of view of the climate change regime is a matter of opinion. On the one hand, it might be considered negative because it restricts the set of policy options open to governments. Environmentalists often argue that because of this restriction governments set less ambitious reduction targets than they would otherwise have done. On the other hand, the discipline of the WTO reduces the temptation of governments to (ab)use climate change policies for strategic, commercial purposes, and it probably strengthens the incentive for governments to co-operate in the implementation of their policies. It is generally acknowledged that the climate change regime can benefit from the discipline that has been developed in the world trade regime

At the material level, there are also potential synergies and conflicts between the climate change regime and the world trade regime. A negative linkage from the point of view of the climate regime is the possibility of carbon leakage. Through the mechanisms of international trade, a unilateral reduction of emissions of greenhouse gases in those countries that have reduction obligations under the Kyoto Protocol (Annex B countries) could result in an increase in the optimal level of greenhouse gas emissions in countries that have no reduction obligations (non-Annex B countries). There is still much uncertainty on the potential magnitude of carbon leakage.

The positive linkage is that adjustments through international trade reduce the compliance costs of Annex B countries. These compliance costs can be reduced even further by the use of specific trade and investment mechanisms such as emissions trading, joint implementation and the clean development mechanism. Lastly, but of no small importance, international trade can play a significant role in the international diffusion of clean technologies.

All in all, there are many linkages between international climate policy and other international issues, environmental or otherwise. In some cases linkages are small, but in others the linkages have substantial effects. International climate policy should therefore not be conducted in isolation, nor should it be researched in isolation.

NOTES

1. Three referees helped improve the exposition. Greenpeace International, the US National Science Foundation through the Center for Integrated Study of the Human Dimensions of Global Change (SBR-9521914), and the Michael Otto Foundation provided welcome financial support. All errors and opinions are ours.
2. Exceptions to the MFN include the preferential agreements concluded with developing countries, and the establishment of free-trade areas such as the EU.
3. Those criteria were first proposed in the report of the 1970 Working Group on Border Tax Adjustments, BSID 18S/97 (1972) at 18.

4. Article XX GATT *General Exceptions*:
   'Subject to the requirement that such measures are not applied in a manner which would constitute a means of arbitrary or unjustifiable discrimination between countries where the same conditions prevail, or a disguised restriction on international trade, nothing in this agreement shall be construed to prevent the adoption or enforcement by any contracting party of measures:
   (b) necessary to protect human, animal or plant life or health
   (g) relating to the conservation of exhaustible natural resources if such measures are made effective in conjunction with restrictions on domestic production or consumption.'
5. *United States – Standards for Reformulated and Conventional Gasoline*, Panel Report adopted on 29 January, 1996, WT/DS2/R, paras. 6.21 and 6.37.
6. *Thailand – Restrictions on Imports and Internal Taxation of Cigarettes*, Panel Report adopted on 7 November, 1989, BSID 37S/200 (1990). See in particular para. 75 'the import restrictions imposed by Thailand could be considered to be necessary in terms of art. XX(b) only if there were no alternative measure consistent with the [GATT] or less inconsistent with it, which Thailand could reasonably be expected to employ to achieve its health policy objectives.'
7. *Canada – Measures Affecting Exports of Unprocessed Herring and Salmon*, panel report adopted on 22 March, 1988, BSID 35S/98 (1988), para. 4.5.
8. *United States – Import Prohibition of Certain Shrimp and Shrimp Products*, AB Report adopted on 12 October, 1998, WT/DS58/AB/R, paras. 136 and 141.
9. *United States – Standards for Reformulated and Conventional Gasoline*, AB report adopted on 29 May, 1996, WT/DS2/AB/R (1996), para. III.C.
10. *United States – Standards for Reformulated and Conventional Gasoline*, AB Report, para. IV. See also *European Communities – Measures Affecting Asbestos and Asbestos-Containing Products*, Panel Report adopted on 18 September, 2000, WT DS135/R. See in particular paras. 8.238-8.239, where the panel ruled that a national measure would qualify as a 'disguised restriction on international trade' if compliance with it would only be a disguise to conceal the achievement of trade-restrictive objectives. In this regard, one should also recall that the panel added that the mere fact that a national measure benefits domestic production did not constitute a protectionist aim as such, provided it remains within certain limits.
11. TBT, Annex I defines 'technical regulation' as a document that lays down product characteristics or their related processes and production methods (...) with which compliance is mandatory. It may also include terminology, symbols, packaging, marking or labelling requirements.
12. *United States – Import Prohibition of Certain Shrimp and Shrimp Products*, Appellate Body Report, WT/DS58/AB/R, circulated on 12 October, 1998, para. 121.
13. *United States – Restrictions on Imports of Tuna* (unadopted), 30 International Legal Materials 1594 (1991) and *United States – Restrictions on Imports of Tuna* (unadopted), 33 International Legal Materials 839 (1994).
14. See *United States – Import Prohibition of Certain Shrimp and Shrimp Products*, WT/DS58/R, para. 61 'do not imply that recourse to unilateral measures is always excluded, particularly after serious attempts have been made to negotiate'.
15. In addition to border tax adjustments, regulatory standards and subsidies, Annex I parties may launch, for example, eco-labelling schemes to promote the sale of

climate-friendly products. Eco-labels may either fall under the TBT or under the GATT. See *United States – Standards for Reformulated and Conventional Gasoline*, WT/DS2/AB/R (1996).

16. One key issue is to determine whether fuel and other energy-related products are 'like' products or not.

17. *United States – Taxes on Petroleum and Certain Imported Substances*, BSID 34S / 136 (1987), para. 5.2.7

18. See *European Communities – Measures Affecting Asbestos and Asbestos-containing products*, AB Report adopted on 12 March, 2001,WT/DS135/AB/R. See in particular paras. 101 et seq., where the AB held that chrysolite asbestos fibres and asbestos-free fibres ought not to have been considered as 'like' products given their respective levels of carcinogenicity, it held that the health risk posed by each of those two products should have been taken into account as an element of their physical characteristics.

19. One should first recall that the exchange of PAAs between Annex I parties should not be covered by the WTO. It should instead be regarded as a state-by-state re-allocation of PAAs, not the creation as such of a market in goods or services. Therefore, any Annex I party shall remain free to decide whether or not to sell PAAs, and to whom they wish to sell them. For more details, see Werksman (1999).

20. Note that the trade in allowances may fall under the General Agreement on Trade and Services. Indeed, an emission permit might be interpreted as a 'negotiable instrument' with monetary value. Those Annex I parties who have adhered to the Financial Sector Binding Schedule under the GATS Agreement on Financial Services would then be obliged to open their markets to trade in emission permits. They would not be able to prevent brokers from buying and selling emission allowances to and from foreign companies. However, parties would have no obligation to recognize as valid the allowances bought and sold, and could still prohibit their use on its domestic territory.

21. The economic impact of limiting the emissions of other greenhouse gases are understood less well. At the time of writing, the role of sinks to sequester CO_2 in the Kyoto Protocol is not very clear. Non-CO_2 greenhouse gases and sinks may affect the economic impact of the Kyoto Protocol in a significant way. It has been estimated that limiting the emissions of non-CO_2 gases may reduce overall costs by two-thirds (Reilly et al., 1999).

22. The elasticity of supply depends, among other things, on OPEC's ability to exercise market power over the supply of oil, such as limiting a reduction in the world price of oil (IPCC, 2001).

23. Burniaux and Oliveira Martins (1999) add that the more price-elastic the supply of fossil fuels (especially coal) is, the less carbon leakage there will be. Different models make different assumptions on the supply elasticities of fossil fuels.

24. Most economic models mentioned in this chapter do not include non-CO_2 greenhouse gases.

25. WorldScan predicts negative effects on the GDP of these countries. This could be due to the regional aggregation used in WorldScan.

REFERENCES

Bhagwati, J.N. (1958), 'Immiserizing Growth: A Geometrical Note', *Review of Economic Studies*, **25**, 201–5.

Bernstein, P.M., W.D. Montgomery, T.F. Rutherford and G.-F. Yang (1999), 'Effects of Restrictions on International Permit Trading: The MS-MRT Model: The Costs of the Kyoto Protocol: A Multi-Model Evaluation', *The Energy Journal* (Special Issue), 221–56.

Bollen, J., T. Manders and P. Tang (2000), *Winners and Losers of Kyoto*, Bilthoven/The Hague: National Institute for Public Health and Environment, and Netherlands Bureau for Economic Policy Analysis.

Burniaux, J.-M. and J. Oliveira Martins (1999), *Carbon Emission Leakage: A General Equilibrium View*, Economics Department Working Paper No. 242, Paris: OECD.

Calster, G.V. (2000), *International & EU Trade Law: the Environmental Challenge*, London: Cameron May International Law & Policy.

Cooper, A., S. Livermoore, V. Rossi, A. Wilson and J. Walker (1999), 'The Economic Implications of Reducing Carbon Emissions: A Cross-Country Quantitative Investigation using the Oxford Global Macroeconomic and Energy Model', in 'The Cost Of The Kyoto Protocol: A Multi-Model Evaluation', *The Energy Journal* (Special Issue), 335–65.

Copeland, B.R and M.S. Taylor (2000), 'Free Trade and Global Warming: A Trade Theory View of the Kyoto Protocol', NBER Working Paper Series, Cambridge, MA: National Bureau of Economic Research.

IPCC (2001), *Climate Change 2001: Mitigation*, A report by the Intergovernmental Panel on Climate Change.

Kainuma, M., Y. Matsuoka and T. Morita (1999), 'Analysis of Post-Kyoto Scenarios: The Asian-Pacific Integrated Model', in 'The Cost of the Kyoto Protocol: A Multi-Model Evaluation', *The Energy Journal* (Special Issue), 207–20.

Manne, A.S and R.G. Richels (1999), 'The Kyoto Protocol: A Cost-effective Strategy for Meeting Environmental Objectives?', in 'The Costs of the Kyoto Protocol: A Multi-Model Evaluation', *The Energy Journal* (Special Issue), 1–23.

McKibbin, W.J., M.T. Ross, R. Shackleton and P.J. Wilcoxen (1999), 'Emissions Trading, Capital Flows and the Kyoto Protocol', in 'The Costs of the Kyoto Protocol: A Multi-Model Evaluation', *The Energy Journal* (Special Issue), 287–333.

OECD (1999), *Action Against Climate Change*, Paris: OECD.

Reilly, J., R. Prinn, J. Harnisch, J. Fitzmaurice, H. Jacoby, D. Kicklighter, P.

Stone, A. Sokolov and C. Wang (1999), 'A multi-gas assessment of the Kyoto Protocol', *Nature*, **401**, 549–55.

Sands, P. (2000), 'Environmental Protection in the Twenty-First Century: Sustainable Development and International Law', in R.L. Revesz, P. Sands and R.B. Stewart (eds), *Environmental Law, the Economy, and Sustainable Development*, Cambridge: Cambridge University Press, pp. 369–409.

Tol, R.S.J., W. Lise, B. Morel and B.C.C. van der Zwaan (2001), *Technology Development and Diffusion and Incentives to Abate Greenhouse Gas Emissions*, Research Unit Sustainability and Global Change SGG-6, Centre for Marine and Climate Research, Hamburg University, Hamburg.

Tulpulé, V., S. Brown, J. Lim, C. Polidano, H. Pant and B.S. Fisher (1999), 'The Kyoto Protocol: An Economic Analysis Using GTEM', in 'The Costs of the Kyoto Protocol: A Multi-Model Evaluation', *The Energy Journal* (Special Issue), 257–85.

Werksman, J. (1999), *Greenhouse Gas Emissions Trading and the WTO*, RECIEL, **8**(3), 251–64.

Weyant, J.P. (ed.) (1999), 'The Costs of the Kyoto Protocol: A Multi-Model Evaluation', *The Energy Journal* (Special Issue), 11–398.

WTO (1996), Report of the (WTO) Committee on Trade and Environment, WT/CTE/1 (96-4808), posted at www.wto.org.

11. Elaborating an international compliance regime under the Kyoto Protocol

Juliette van der Jagt[1]

1 INTRODUCTION

In general, it is difficult to establish the means to ensure compliance with the obligations of international law and, more especially, to respond to non-compliance. Because international law is essentially a regime of co-ordination, it lacks a centrally regulated enforcement system to identify and react to cases of non-compliance with its obligations. The power of the organs of international organizations to enforce decisions depends on the consent of the parties to a treaty. Such consent is often difficult to secure since states are generally unwilling to submit to strong enforcement mechanisms or to transfer too much enforcement power to international institutions (Sands, 1996). Instead, they want to retain their sovereign powers.

Nevertheless, it is of the utmost importance that the rules of international law are accompanied by instruments to ensure compliance and to respond to non-compliance. Compliance control is an inherent feature of international society. If there is no response to non-compliance with the provisions of a treaty the overall effectiveness of the treaty will be limited, and the commitments which have been made under the international legal process will be undermined (for example Sands, 1996). Furthermore, compliance control may lead states genuinely to observe the rules, thereby further protecting the environment.

Compliance control is important in all fields of international law, but particularly in international environmental law. One reason for this is that the increasing costs of compliance with environmental regulations means that states have a greater interest in ensuring that other states subject to the same international regulations also live up to their obligations, thereby ensuring competition on a level playing field (Handl, 1997). Particularly in relation to climate change, at least two further reasons can be added: the enormous impact measures relating to climate change will have on national economies and the prevention of the environmental damage resulting from climate change which may particularly affect certain states.

Since the adoption of the Kyoto Protocol, the design of a compliance system has been one of the most important issues. The Fourth Conference of the Parties (COP-4) established the Joint Working Group on Compliance (JWG) under the Subsidiary Body for Implementation (SBI) and the Subsidiary Body for Scientific and Technological Advice (SBSTA) in Decision 8/CP.4. The JWG was assigned the extremely difficult task of developing a comprehensive compliance system. The elements of the compliance regime can be found in Articles 5, 7, 8 and 18 of the Protocol. The system will apply specifically and solely to the obligations of the Kyoto Protocol, thus not to any other areas of international environmental law. It will be an internal procedure, regulating the control of parties' compliance with the Protocol by an international institution functioning under the treaty.

The aim of this chapter is to describe and analyse how the compliance regime under Article 18 of the Kyoto Protocol should and, according to the Marrakesh Accords adopted at the seventh session of the COP (COP 7), will be structured. To this end, the chapter has the following structure. Section 2 will briefly explain the terms 'compliance' and 'enforcement' which play an important role in the discussion. Section 3 will discuss the various parts of the Kyoto compliance regime in general terms. Because this contribution focuses on Article 18 as the core provision of the Protocol dealing with (non-)compliance, the procedures under Articles 5, 7 and 8 will be only briefly discussed. It must be stressed, however, that these procedures will serve as the basis for the procedure that is designed to determine and address cases of individual non-compliance and, therefore, will play an important role. Section 4 will then discuss more concretely some of the procedures and mechanisms that should and some of which will form part of the compliance system based upon Article 18 of the Protocol. The chapter ends with some conclusions in Section 5.

This chapter will only touch upon compliance provisions relating to the use of the flexible mechanisms. For a discussion of the role that these mechanisms will play in meeting the quantitative emission commitments, see Chapter 7. Finally neither traditional and external forms of international enforcement, such as dispute-settlement procedures, nor the possibilities for individuals and non-governmental organizations (NGOs) to 'enforce' treaties are considered here.

2   SOME DEFINITIONS

To avoid any confusion, for the purpose of this contribution two terms need to be distinguished and defined: 'compliance' and 'enforcement'.

Compliance can be defined as the extent to which the behaviour of a party to a treaty actually meets its obligations as laid down in that treaty. In practice, the implementation of the commitments entered into by the state requires actions to be taken at both the national and the international level (Wolfrum, 1999). Compliance is the result of implementation and enforcement. Compliance systems include measures to prevent non-compliance, to determine (non-)compliance, to facilitate compliance and to address cases of non-compliance. Under the Kyoto Protocol these functions are fulfilled by procedures included in or based on Articles 5, 7, 8 and 18. These procedures are used to ensure that the behaviour of a party indeed corresponds to its treaty obligations and hence ensure compliance with the obligations of the Kyoto Protocol.

Enforcement is the set of actions to be taken in response to an identified case of non-compliance in order to bring the violator back to compliance. In this sense, enforcement has a narrow meaning: it is one of the components of the compliance regime that parties to a treaty may adopt and it refers to the negative responses (that is sanctions) to cases of non-compliance (cf. Wolfrum, 1999). Positive responses to (possible) non-compliance are generally referred to as facilitation responses. These are not covered by the term 'enforcement'. In relation to the climate change regime, responses, either facilitating or sanctioning, are also referred to as 'consequences'. As this is not common legal terminology, this term will not be used in this contribution.

It must be borne in mind that enforcement can also be understood in a broad sense, referring to both responses to non-compliance and supervision of the treaty provisions. Supervision may consist of monitoring, reporting, verification and review. It envisages means to prevent non-compliance and to determine (non-)compliance. As a result, the term 'enforcement' encompasses not only mechanisms to address cases of non-compliance but also mechanisms to determine (non-)compliance and to prevent non-compliance. For the most part, I refer to enforcement in this sense in Section 1. A synonym for this broad meaning of enforcement is the term 'compliance control', which is also mentioned in that section.

3   THE DIFFERENT PARTS OF THE KYOTO
       COMPLIANCE PROCEDURE

Compliance procedures are relatively new in international law and supplement the procedures that apply under the rules of general international law (such as dispute-settlement procedures). Their purpose is not so much to designate the acts of a party as wrong and to impose sanctions, but more to assist violating parties to return to compliance. This implies that a compliance procedure looks forwards rather than backwards (Werksman, 1996). Furthermore, compliance procedures focus on confidence building and co-operation between the parties to a treaty rather than on authoritative or confrontational means (Marauhn, 1996). A compliance regime therefore not only prescribes the methods that can be used to impose sanctions in the event of non-compliance, but also prescribes the means that can be applied to avoid violations and to facilitate compliance.

Regimes designed to ensure that treaty obligations are complied with may comprise three separate though interrelated and sometimes overlapping parts (cf. Corfee Morlot, 1998; Nollkaemper, 1993). During the first part, the international organ concerned has to gather information on the degree to which states comply with their obligations. The purpose of *monitoring* and *reporting* is to gather the information needed to enable compliance assessment. These are important elements, because, to a great extent, they make it possible to prevent (serious) violations. This is of great significance in environmental law where the goal is to ensure compliance before the environment is harmed. Subsequently, during the second part the organ in question has to review to what extent states comply with their international obligations. *Review* and *verification* use the reported information to ascertain compliance. The elements of the first and second part function as supervision mechanisms, and are consequently mechanisms for preventing non-compliance and which form the basis for determining non-compliance. Finally, if it has been determined that a state has failed to comply with an international obligation, it has to be decided whether and to what extent there should be a response to this non-compliance (*facilitation* and *enforcement*). As will be explained below, facilitation does not necessarily have to be a reaction to non-compliance.

These elements of a compliance regime can also be found in the Kyoto Protocol, although they do not always explicitly refer to (non-)compliance. The procedures laid down in Articles 5, 7, 8 and 18 will have to be seen in connection with each other.

Article 5 requires Annex I parties to have in place 'a national system for the estimation' of greenhouse gas emissions by sources and sinks covered by the Protocol at least one year prior to the beginning of the first commitment period (that is by 2007).

According to Article 7, Annex I parties will have to incorporate in their annual inventories and national communications the supplementary information needed to ensure/demonstrate compliance with the Kyoto Protocol (national reporting).

National reports are the main source of information for the verification process. According to Article 8, expert review teams will review the information submitted under Article 7 as part of the annual compilation and accounting of emissions inventories and assigned amounts respectively, as part of the review of communications. Article 8 builds upon and strengthens the review process established by the UNFCCC, experience with which has been positive. The expert review teams are directed to conduct 'a thorough and comprehensive technical assessment of all aspects of the implementation by a Party of this Protocol' and to prepare a report for the Conference of the Parties serving as the meeting of the parties (COP/MOP) at which 'the implementation of the commitments of the Party' is assessed and 'any potential problems in, and factors influencing, the fulfilment of commitments' are identified (Article 8.3). The Secretariat will circulate the reports to all parties to the convention and will forward any 'questions of implementation' indicated in the expert teams' reports for consideration by the COP/MOP (Article 8.3). In this way, factors and problems affecting compliance can be communicated to the parties.

Article 18 mandates the COP/MOP to approve 'appropriate and effective procedures and mechanisms to determine and to address cases of non-compliance with the provision of this Protocol, including through the development of an indicative list of consequences'. This article focuses on facilitation and enforcement, which will be explained below.

The aim of facilitating compliance – also referred to as 'compliance assistance' – is to improve the ability of states to meet their treaty obligations. This facilitative approach helps to ensure that all parties have the institutional, technical, and financial capacity to fulfil their obligations under the treaty. Furthermore, it has the advantage of encouraging parties to come forward and notify the compliance authority of any difficulties they are experiencing in complying with their obligations. This implies that facilitation can be carried out even where the party concerned may not have formally been found to be in non-compliance. Compliance assistance may be useful if parties lack the capacity to comply (Brown-Weiss, 1997; Chayes and Chayes, 1995). Where the problem of non-compliance is due to a lack of capacity, there is little point in resorting to sanctions.

The assistance that the facilitation approach may provide is particularly relevant for multilateral environmental agreements, such as the Kyoto Protocol, in which the parties must deploy new technologies or make fundamental institutional changes in order to meet their obligations. Furthermore, facilitation will confirm or strengthen the confidence which developing states may have in the climate regime. Therefore, facilitation responses will play a key role in the Kyoto compliance system (Goldberg et al., 1998; Corfee Morlot, 1998; Werksman, 1998).

It must be stressed that the reason for providing assistance is not necessarily related to compliance at all; assistance could also be a preventive means of directing state behaviour if integrated into the treaty from the outset (Beyerlin and Marauhn, 1997). Articles 10 and 11 of the Kyoto Protocol are examples of this approach.

In international environmental law, enforcement mechanisms (sanctions) are rarely used; they are often merely seen as an 'ultimum remedium', as 'a last resort after other methods have failed' (Brown-Weiss, 1997). However, the inclusion of negative responses to non-compliance will be essential in ensuring observance by the parties to international obligations. Although compliance systems are co-operative and non-adversarial, there is a growing feeling that it should be possible to actually impose sanctions where necessary, for instance when a party persists in violation (for example Marauhn, 1996; O'Connell, 1995; Corfee Morlot, 1998). If a compliance system is to be effective and contribute to environmental protection (and in general it will do both) it will have to contain sanctions and they must be imposed in some cases. If not, the compliance regime will lose credibility. Of course, since the environmental damage caused by non-compliance must be prevented, parties should in the first place be deterred from not complying. But if this fails a sanction will have to be applied (except when it is more appropriate to use a facilitation response). The effect of the compliance system will then be either to repair the effects of non-compliance or to punish the violator.

The possibility of imposing sanctions is particularly significant with respect to the Kyoto Protocol. Climate change due to global warming poses a profound threat to humankind's well-being. Taking meaningful action to counter the threat calls for a substantial change in state behaviour and may impose significant costs on state economies. For these reasons, effective implementation of the Protocol requires not only the availability of facilitation measures but also of enforcement responses, so that it is possible to respond to the many different types of compliance problems that could arise (Wiser, 1999). Furthermore, according to some scholars 'the demand for and use of enforcement approaches may rise as treaties attempt to achieve deeper international co-operation on the environment' (Raustiala and Victor,

1998). This implies that the dependence of parties on each other's compliance increases, which is the case in relation to the UNFCCC and the Kyoto Protocol (Werksman, 1998). All this makes the demand for enforcement responses far higher under the Kyoto Protocol than under other multilateral environmental agreements (Oberthür and Ott, 1999). And these responses are not to be found in those treaties. A growing interest in tougher responses to non-compliance was also reflected in the negotiations on the Kyoto Protocol (cf. Werksman, 1998). In addition, according to literature published before the adoption of the Protocol, the aspect of enforcement should play an important role in the Protocol (for example Addink, 1997).

The procedures under Article 18 concern not only responses but also aspects such as the structure of the compliance system and the procedures to be followed. These aspects and the possible responses to non-compliance in the climate regime will be discussed in the next section.

4 PROCEDURES AND MECHANISMS RELATING TO COMPLIANCE UNDER ARTICLE 18 OF THE KYOTO PROTOCOL

During the work of the JWG an increasing number of elements were recognized as being important components of the compliance system. Among the elements that needed to be included in the compliance regime were: the form of the compliance procedure; the objective of the compliance system; the guiding principles for the operation of the system; the establishment and structure of the compliance body; the submission of questions of implementation; preliminary procedures; proceedings of the compliance body; appeal; the role of the COP/MOP; an additional period for fulfilling commitments; responses to (potential) non-compliance; and the relationship with Articles 16 and 19 of the Protocol (cf. FCCC/SB/2000/CRP.15/Rev.2).

At its resumed sixth session, the COP adopted the Bonn Agreement for the Implementation of the Buenos Aires Action Plan (hereinafter the 'Bonn Agreement') (Decision 5/CP.6, FCCC/CP/2001/L.7). This Agreement contains binding decisions on compliance. After the adoption of the Bonn Agreement, many details on compliance remained to be decided. Decisions in this respect have been taken at COP-7 in the Marrakesh Accords (Decision 24/CP.7, FCCC/CP/2001/13/Add. 3). At the latest COP, COP-8, decisions have not been taken on compliance issues.

Some of the elements mentioned above will be analysed in the following subsections. I will indicate how these elements should in my view be elaborated, in most cases supported by views of other scholars and conference papers. Furthermore, in relation to each element I will discuss the decision of COP-7. Because of the limited space available, it is impossible to discuss the views of other scholars and all aspects of the elements in great detail here.

4.1 Form of the Compliance System

An important question is what form of compliance system should be adopted. The second sentence of Article 18 explicitly states that procedures and mechanisms 'entailing binding consequences shall be adopted by means of an *amendment* to this Protocol' (emphasis added). Thus, at least when the responses are binding, the form in which the compliance regime will be adopted is clear. As a purely legal matter a 'binding consequence' is a 'consequence that affects a party in a compulsory manner or places it under a definite duty or legal obligation to do something it would otherwise not be obliged to do' (Goldberg et al., 1998, referring to *Black's Law Dictionary*, 5th edition, 1979, 153). Thus, the response will lead to a result that is additional to the party's original obligation. The amendment procedure, however, raises some problems that are difficult to overcome. For instance, an amendment will only enter into force when it has been expressly accepted by three-fourths of the parties to the Protocol and the amendment will only bind the parties that have expressly accepted it (Article 20 of the Protocol). This implies a possibly uneven application of the compliance system between the parties to the Protocol (Oberthür and Ott, 1999). These problems may be overcome if the compliance procedure is adopted by a *decision* of the COP/MOP. An important advantage of this approach is that it assumes validity for all parties to the Protocol upon adoption (cf. Ott, 2001). This approach has also been used with regard to the compliance procedure of other treaties, for example the Montreal Protocol. However, it is questionable whether a decision of the COP/MOP could be a sound legal basis for adopting legally binding responses (Werksman, 1998).

During COP-6, Part 2, the issue nearly deadlocked the negotiations and it appears that it has not been clearly resolved. Some parties did not want the compliance regime to have a legally binding nature, whereas other parties argued that in order to be effective the Protocol requires legally binding compliance provisions and sanctions. Because of this conflict, the COP only agreed to adopt the compliance procedures and responses to non-compliance as specified in its decision, and to recommend to the COP/MOP that it should adopt procedures and mechanisms relating to compliance in terms of Article 18 of the Kyoto Protocol. This decision has been affirmed at COP-7. It was thus not made explicit through which instrument the system will be adopted. Considering the responses that the COP agreed upon (which are partly legally binding), this is not a very consistent decision.

4.2 Establishment and Structure of the Compliance Committee

Although the Protocol already provides for several organs, an institutional arrangement that is to play the central role in the compliance system should be established. This new institution should be a standing body; the nature and breadth of its duties and the importance of consistency in its work require this. The existence of a compliance body will be crucial for the adequate functioning of the compliance regime.

The Marrakesh Accord indeed establishes such a compliance body, called the Compliance Committee. The functions of this body were a topic of much debate among the parties. To resolve the issue, the COP decided that the Compliance Committee will consist of two governing bodies or 'branches', a facilitative branch (FB) and an enforcement branch (EB). Furthermore, the COP took decisions on the mandate of these branches, an issue which had also been a bottleneck. Generally, the FB shall be responsible for providing advice and facilitation to parties in implementing the Protocol, and for promoting compliance by parties with their commitments under the Protocol. Especially, the FB will be responsible for addressing questions of implementation relating to Article 3.14 and with respect to the provision of information on the use by an Annex I party of the Articles 6, 12 and 17. Furthermore, the FB shall be responsible for providing advice and facilitation for compliance in relation to the quantitative emissions commitments (Article 3.1) and methodological and reporting requirements (Articles 5.1, 5.2, 7.1 and 7.4). The EB shall be responsible for determining whether an Annex I party is not in compliance with the quantitative emission commitments, monitoring and reporting requirements and eligibility requirements under the flexible mechanisms (Articles 6, 12 and 17). The EB will serve as a quasi-judicial forum. The mandate relating to the monitoring and reporting requirements (Articles 5 and 7) makes clear that these requirements play a role not only in the compliance regime, but that these are first and foremost primary conditions which the parties must fulfil in order to be able to join the Kyoto regime in the first place. Therefore, it is important to ensure compliance with these Articles and to respond to cases of non-compliance. COP-7 has strengthened this by the decision to recommend to the COP/MOP that the eligibility to participate in the mechanisms by an Annex I party will depend on its compliance with methodological and reporting requirements. Thus, being in compliance with the procedural obligations is a precondition for using the flexible mechanisms.

It is indispensable for the successful functioning of the compliance system that the branches are independent and impartial. The members should therefore act in their personal capacity and not as representatives of the parties. This applies particularly to the EB, which should not be under any pressure to react to political concerns. In this way, the principle of due process is taken into consideration (CIEL/WWF, 2000). The issue of the size and composition of the compliance bodies, also the subject of fierce debate among the parties, was also resolved at COP-7: the FB and the EB will each be composed of 10 members, one member from each of the five regional groups of the United Nations and one member from the group of AOSIS; two members from Annex I parties and two members from non-Annex I parties. It is fortunate that the COP has established a body with a limited membership since a smaller body will be able to take action more easily in response to a situation of non-compliance than a body consisting of many members.

Bearing in mind the need for different approaches to non-compliance (cf. Section 3), the COP decided with good reason to split up the Compliance Committee into two bodies of a different nature. As explained in Section 3, the Kyoto compliance procedure cannot possibly function without an enforcement mechanism. Meanwhile, it is the first time in 'environmental law history' that a compliance procedure has such a dual nature. Facilitation has been included in previous multilateral environmental agreements (for example the Montreal Protocol), but enforcement as provided for in the climate change regime has not.

4.3   Initiating the Compliance Procedure

The next issue to be addressed involves who should have the competence to initiate the procedure before the branches. The reports of the expert review teams (Article 8) will play an important role here. Obviously, the Secretariat should have the competence to initiate the procedure by submitting these reports, including the questions of implementation indicated in the reports and listed by the Secretariat, to the Compliance Committee (cf. CIEL/WWF, 2000).

The Secretariat will be in a good position to initiate proceedings since it oversees the implementation of the Kyoto Protocol and has the necessary skilled personnel. It is also independent and impartial. To some extent, Article 8.3 already provides the legal basis for this function of the Secretariat (Oberthür and Ott, 1999). Furthermore, the Secretariat may also be empowered to raise compliance issues not included in the expert review reports. For example, compliance issues relating to non-Annex I parties (developing countries) will not be included in these reports because they are not subject to the Article 8 review process. According to Article 12 of the UNFCCC, however, these countries also have to submit reports to the Secretariat. The Secretariat may identify compliance issues arising from the reports and duly forward them to the Compliance Committee.

In addition to the Secretariat, parties should be competent to trigger the procedure with respect to the implementation of their own obligations and with respect to the implementation of another party's obligations. At COP-7 it has been decided that the compliance procedure can be initiated by reports of the expert review teams, the Secretariat or the parties. It remains to be seen whether a party will, in practice, trigger the compliance procedure concerning the potential non-compliance of another party. States might be unwilling to initiate procedures against other states for the fear of retaliation. Therefore, it is not certain whether parties will initiate procedures against other non-complying parties. However, given the differentiation of commitments under the Protocol, developing countries may more readily trigger the compliance procedure against developed states (Werksman, 1996).

4.4   Preliminary Procedures

Before the substance of the case can be dealt with it has to be assigned to the appropriate branch. It has to be decided which branch will deal with the case. According to the Marrakesh Accord, this decision will be taken by the bureau of the Committee, being the chairpersons and vice-chairpersons of the branches. The allocation revolves around the administrative reference of the case. This administrative reference will have significant consequences because at this stage the decision will in principle be taken on whether facilitation or enforcement will follow.

After the case has been submitted to the relevant branch there will be a preliminary examination. This formal part will precede the substantive part of the case. COP-7 has stated that this examination will be undertaken to ensure that the case is supported by sufficient information, is not *de minimis* or ill-founded and is based on the requirements of the Protocol. It is logical that the branch itself and not another organ (cf. 'the bureau') will carry out this examination. The same organ should be competent with respect to both the formal and the substantive part as this provides transparency in the compliance procedure.

The preliminary examination will end in the decision to proceed or not to proceed.

4.5   Proceedings at the Branches

If after the preliminary examination the relevant branch has decided to proceed, it will discuss the case. The branch will have to determine whether or not the party can be found to be in non-compliance. In order to carry out a compliance assessment the relevant branch should use all the relevant information it can obtain. This information could be provided by reports from the expert review teams, the party concerned, reports from the COP, COP/MOP, the subsidiary bodies and the other branch. Furthermore, information could be submitted by any intergovernmental organizations and NGOs that have relevant factual and technical information. But in my opinion not all NGOs should be free to submit information in all cases as this may cause both practical and political problems. Instead, only certain NGOs, for example NGOs which are included in 'a list of competent organizations', should be requested to submit information (for the opposite case see for example CIEL/WWF, 2000). Finally, expert advice may be used. According to the Marrakesh Accord, the branches may use all these sources of information to base their deliberations on. The COP did not restrict the role of the NGOs in any sense.

The procedural arrangements will be more extensive in respect of the enforcement procedure than for the facilitation procedure as the former will in principle result in more weighty consequences for the party concerned. Therefore, apart from some general rules applicable to both procedures (such as the representation of the parties, sources of information and translation), COP-7 created special arrangements in relation to the enforcement procedure. These include, among other things, the possibility for the party concerned to make written submissions, the right to a hearing and the possibility to present expert testimony or opinion at the hearing. Furthermore, the COP decided that the EB needs to adopt a preliminary finding that the party concerned is not in compliance with its commitments before it takes a final decision (unless the EB determines not to proceed with the case). In my opinion, the EB does not need to make and issue a preliminary finding before it takes a final decision as the compliance procedure will also provide the possibility of an appeal (cf. Section 4.8, below). The provision for both a preliminary finding and an appeal (as well as a preliminary examination and a final decision) implies that the enforcement proceeding could drag on for years, which should of course be avoided. Nor is it necessary to provide both possibilities as due process can already be ensured by one of these procedures. Therefore, it is regrettable that the COP decided otherwise. Finally, the COP took a special decision in relation to voting by the EB. Its decisions will require a majority of members from Annex I parties as well as a majority of members from non-Annex I parties.

4.6 Responses to Non-compliance

The proceedings of the facilitation and enforcement branches can logically result in three types of decision. The branch can decide that the case has been resolved, which implies that there is no longer a question of non-compliance. The branch concerned may also refer the case to the other branch if it feels that branch is better equipped to decide on it. The latter should then deal with the case. Finally, the branch may decide to invoke a response against the party concerned.

Quite a few proposals have been made in conference papers and in literature concerning the measures that the FB and EB may take. During COP-7, the COP agreed on both facilitation and enforcement responses which will be discussed below. The facilitation responses are less specific than the enforcement responses though.

Facilitative responses will mainly consist of some kind of assistance. The FB will be competent to provide mechanisms for financial assistance. The funding for climate change is provided through the Global Environmental Facility (GEF) on the basis of Article 4.3 UNFCCC. Facilitating compliance may also involve the facilitation of technical assistance, including technology transfer and capacity building. Financing and technology transfer reflect the principle of common but differentiated responsibilities (see Article 3 UNFCCC). Furthermore, the FB will have the competence to make recommendations relating to actions to be taken by the party concerned in implementing the Protocol. Finally, it was decided at COP-7, that the FB will provide advice to individual parties regarding the implementation of the Kyoto Protocol.

According to the Marrakesh Accord, the FB will provide advice and facilitation with the aim of promoting compliance and providing early warning of any potential non-compliance. This implies that the FB will function without the need for a party to be actually found to be in non-compliance. Therefore, it will not consider such cases only *in response* to a case of non-compliance.

The responses which the EB can impose are related to the provisions of the Kyoto Protocol which have been violated. As far as violation with Article 3.1 is concerned, the Marrakesh Accord establishes that the responses for non-compliance with this Article will be the subtraction of excess emissions from assigned amounts, the development of a compliance action plan and the suspension of the eligibility to trade under Article 17. According to the COP, these responses will be aimed at the termination of non-compliance in order to ensure environmental integrity and providing an incentive to comply.

As regards the response of subtraction of excess emissions, it has been decided that for every tonne of emissions by which a party exceeds its target during the first commitment period, 1.3 tonnes will be deducted from its assigned amount for the subsequent commitment period. Thus, the amount subtracted will equal the excess multiplied by a penalty rate (that is 0.3). The rate will serve as an interest rate for the delays in the achievement of emissions reduction. It should also be an incentive to comply and must, therefore, be set at a relatively high level. If it turns out that the rate is not high enough it should be increased for subsequent commitment periods. According to the Marrakesh Accord, for subsequent commitment periods the rate will be determined in future amendments, so it is indeed possible to decide on higher rates. Subtraction and penalty rates will in principle guarantee environmental integrity at least, provided that the emission commitments for the subsequent periods are adopted and enter into force in time. If not, parties may anticipate their being in non-compliance at the end of the commitment period and pursue lower targets for the subsequent commitment period. Subtraction and penalty rates will then not heal the environment. So it is crucial that the targets of the second commitment period are adopted before the compliance regime can be fully developed.

The development of a compliance action plan implies that the non-complying party will, within a specific period of time, develop and submit to the EB a clear and detailed plan demonstrating how it proposes to comply with its quantitative emissions commitment in the subsequent commitment period (cf. Wiser and Goldberg, 2000). The COP decided that the EB will have the power to review the plan and assess whether or not it is likely to work. Furthermore, it agreed that in the compliance action plan the party concerned will have to give priority to domestic policies and measures. Otherwise, it will become even more difficult to meet the target in the second and subsequent commitment periods. The plan will make the party specifically aware of its own failure and of the way it can prove its implementation. It must be doubted, however, whether this will be an effective sanction, especially if the party is not interested in complying. The possible responses if the party in question does not draw up such a plan or fails to implement (parts of) the plan have not been specified. However, at a later stage it might be possible to agree on sanctions in such cases. Anyhow, it is worth envisaging how this sanction might work.

The flexible mechanisms introduce a new set of tools by which to respond to cases of non-compliance. One of these responses is the suspension of the eligibility to make transfers under Article 17. The Marrakesh Accord limits the competence of suspending the eligibility to the *transfer* of tonnes in order to enable Annex I parties still to *acquire* tonnes through using the flexible mechanisms. In this way, these parties can return to the allowed level of emissions as soon as possible. The EB will suspend transfers for a period of time, for instance until the party has demonstrated to the satisfaction of the EB that it will meet its target in the subsequent commitment period. The suspension of a party's right to make transfers is an example of suspending a treaty privilege. Because this restriction will affect privileges created by the Protocol itself and not fundamental rights that sovereign states enjoy outside of the treaty regime, unlike the subtraction of excess emissions and the development of a compliance action plan it is not a 'binding consequence' (cf. Section 4.1, above; Goldberg et al., 1998). As it will have important economic consequences for the party and could result in a party losing face, this sanction will be an incentive to comply.

The fact that the COP has agreed on responses to be applied by the EB is a major breakthrough in international environmental law. The compliance procedure of the Kyoto Protocol will be the first procedure with 'binding consequences'. In fact, the whole enforcement element is an innovation in international law (cf. Chapter 4, above). However, the work has not yet been completed. Other responses should be adopted as well. One of these is the publication of non-compliance with Article 3.1. Publishing details of non-compliance enables other states, NGOs and the 'public' to bring pressure upon governments to bring about changes in their state behaviour. Furthermore, it may deter parties from non-compliance in the first place (Wiser, 1999). The effect of this sanction must not be underestimated and should definitely be one of the sanctions which may be imposed on the party not complying.

4.7 Appeal

Once the enforcement branch has taken a final decision relating to Article 3.1 the case might come to an end. Another possibility is that the party concerned lodges an appeal against this final decision if it believes it has been denied due process. An appeal procedure is a safeguard for obtaining this due process. Nevertheless, for a long time it was uncertain whether such a procedure would be established. But the Marrakesh Accord did establish an appeal procedure. Parties will have the right to appeal against final decisions of the EB, but no appeal will be possible against decisions of the FB. Given the nature of the facilitation procedure it is also unnecessary.

An important question is who should be the appellate body. Originally two possible bodies were mentioned (cf. FCCC/SB/2000/CRP.15/Rev.2): the COP/MOP and an appellate body made up of three persons with relevant expertise. Some parties wanted assurances that decisions of the EB could not be made completely independently of COP/MOP review, and therefore preferred the COP/MOP as the competent body. It could be argued, however, that this option is not in accordance with the principle of due process. The appeal procedure should be independent and impartial, which means that a body with those qualities should be empowered to carry out the procedure. Therefore, the competent body should not be the COP/MOP (a political, policy-making body), but a new – standing – body. Although this implies that yet another institution would have to be created in order to deal with compliance, the requirements just mentioned are even more pertinent. It is, therefore, regrettable that the Marrakesh Accord designates the COP/MOP as the competent appellate body. The decision of the EB will stand unless a majority of at least three-fourths of the COP/MOP vote to overturn it.

5   CONCLUSIONS

The Kyoto Protocol contains the necessary elements to establish a comprehensive compliance system: monitoring is provided for in Article 5, reporting in Article 7, review and verification in Article 8, while Article 18 forms the basis for facilitation and enforcement. Article 18 is the focus of this contribution. Compliance is defined here as the extent to which the behaviour of a party actually corresponds to its treaty obligations. Enforcement is the means by which compliance can be achieved and refers to the negative responses to non-compliance. For their part, facilitation responses envisage positive responses to cases of non-compliance.

To be able to react effectively to cases of non-compliance, the compliance regime under Article 18 should at least contain procedures and mechanisms relating to the form of the compliance system, the establishment and structure of the compliance body, an additional period for fulfilling commitments, the initiation of the procedure, preliminary procedures, the proceedings at the branches, facilitation and enforcement responses and the possibility of appeal. These elements prescribe which steps have to be taken in order to determine and respond to cases of non-compliance, state who has what tasks and clarify that the competent institutions (the compliance bodies, the appellate body) will, and also must, function independently and impartially.

At COP-7, agreement was reached on compliance rules. The Marrakesh Accord establishes a Compliance Committee, a facilitation branch and an enforcement branch. It includes the composition, the mandates and voting procedure of these bodies. Furthermore, the agreement establishes responses to non-compliance (subtraction of excess emissions, development of a compliance action plan and suspension of the eligibility to make transfers) and provides for appeal. However, the issue of the legally binding nature of the compliance regime has not been clearly resolved. It appears that the COP/MOP will have to take decisions on this. It can already be said, though, that the Kyoto Protocol's compliance procedure is the most robust ever adopted for a multilateral environmental agreement. Furthermore, the enforcement element included in the Marrakesh Accord is an innovation in international environmental law.

The compliance system is likely to positively influence the extent to which the norms of the Kyoto Protocol will be complied with and will therefore contribute to the ultimate objective of Article 2 of the UNFCCC. It can function for several commitment periods. Although it obviously depends on how the obligations of the Protocol are fulfilled in practice and the adoption of entirely new obligations for the parties, in theory the compliance system is not an aspect of the Protocol that will require adjustment during every budget period. However, at this moment in time, it cannot but be concluded that only time will tell whether the compliance regime will indeed be able to react adequately to all the uncertainties which the climate regime involves.

NOTE

1. The author would like to thank G.H. Addink, A.H.A. Soons, J. Werksman and J. Cozijnsen for their comments on earlier versions of this contribution.

REFERENCES

Addink, G.H. (1997), 'Norms and Enforcement of the Climate Change Convention. Aspects of Dutch Policy and Some Suggestions about this Meaningful Convention', in Environment Agency of Japan et al. *Climate Change and the Future of Mankind. Pre-COP3 International Symposium on Legal Strategies to Prevent Climate Change* (September 13–14, 1997), Tokyo, 1–14.

Beyerlin, U. and T. Marauhn (1997), *Law-Making and Law-Enforcement in International Environmental Law after the 1992 Rio Conference*, Berlin: Erich Schmidt Verlag.

Brown-Weiss, E. (1997), 'Strengthening National Compliance with International Environmental Agreements', *Environmental Policy & Law*, **27**, 1–9.

CIEL/WWF (2000), *Analysis of Text by the Co-Chairmen of the Joint Working Group on Compliance Dated 10 September 2000, FCCC/SB/ 2000/CRP.7*, Washington: CIEL/WWF.

Chayes, A. and A. Handler Chayes (1995), *The New Sovereignty. Compliance with International Regulatory Agreements*, London, UK and Cambridge/Massachusetts, US: Harvard University Press.

Corfee Morlot, J. (1998), *Ensuring Compliance with a Global Climate Change Agreement*, OECD Information Paper (ENV/EPOC(98)5/REV1), Paris: OECD.

Cozijnsen, C.J.H. and G.H. Addink (1998a), 'Het Kyoto-Protocol onder het Klimaatverdrag: over de inhoud, de uitvoering en de handhaving van afspraken', *Milieu en Recht*, **6**, 152–9.

Cozijnsen, C.J.H. and G.H. Addink (1998b), 'The Kyoto Protocol under the Climate Convention: commitments and compliance', *Change*, **43**, 5–8.

Goldberg, D.M. et al. (1998), *Building a Compliance Regime under the Kyoto Protocol*, Washington/Lisboa: CIEL/EURONATURA.

Handl, G. (1997), 'Compliance Control Mechanisms and International Environmental Obligations', *Tulane Journal of International and Comparative Law*, **5**, 29–49.

Marauhn, T. (1996), 'Towards a Procedural Law of Compliance Control in International Environmental Relations', *Zeitschrift für ausländisches öffentliches Recht und Völkerrecht*, **3**, 696–731.

Nollkaemper, P.A. (1993), *The Legal Regime for Transboundary Water Pollution: Between Discretion and Constraint*, Dordrecht: Martinus Nijhoff Publishers.

O'Connell, M.E. (1995), 'Enforcement and the Success of International Environmental Law', *Indiana Journal of Global Legal Studies*, **3**, 47–64.

Oberthür, S. and H.E. Ott (1999), *The Kyoto Protocol. International Climate Policy for the 21st Century*, Berlin/Heidelberg: Springer-Verlag.

Ott, H.E. (2001), *The Bonn Agreement to the Kyoto Protocol – Paving the Way for Ratification*, Wuppertal: Wuppertal Institute for Climate, Environment and Energy.

Raustiala, K. and D.G. Victor (1998), 'Conclusions', in D.G. Victor, K. Raustiala and E.B. Skolnikoff (eds.), *The Implementation and Effectiveness of International Environmental Commitments: Theory and Practice*, Cambridge: MIT Press.

Sands, P. (1996), 'Compliance with International Environmental Obligations: Existing International Legal Arrangements', in J. Cameron, J. Werksman and P. Roderick (eds), *Improving Compliance with International Environmental Law*, London: Earthscan Publications Ltd., pp. 48–81.

Werksman, J. (1996), 'Designing a Compliance System for the UN Framework Convention on Climate Change', in J. Cameron, J. Werksman and P. Roderick (eds), *Improving Compliance with International Environmental Law*, London: Earthscan Publications Ltd., pp. 85–112.

Werksman, J. (1998), 'Compliance and the Kyoto Protocol: Building a Backbone into a "Flexible" Regime', *Yearbook of International Environmental Law*, **9**, 48–101.

Werksman, J. (1999), *Responding to Non-Compliance under the Climate Change Regime*, OECD Information Paper (ENV/EPOC(99)21/FINAL), Paris: OECD.

Wiser, G.M. (1999), *Compliance Systems under Multilateral Agreements. A Survey for the Benefit of Kyoto Protocol Policy Makers*, Discussion Draft, Washington: CIEL.

Wiser, G.M. and D.M. Goldberg (2000), *Restoring the Balance. Using Remedial Measures to Avoid and Cure Non-Compliance under the Kyoto Protocol*, Washington: WWF/CIEL.

Wolfrum, R. (1999), 'Means of Ensuring Compliance with and Enforcement of International Environmental Law', *Académie de droit international de La Haye*, The Hague/Boston/London: Martinus Nijhoff Publishers, pp. 9–154.

12. Between 'curbing the trends' and 'business-as-usual': NGOs in international climate change policies

Bas Arts and Jos Cozijnsen

---

1 INTRODUCTION

This chapter goes into the role of non-governmental organizations (NGOs) in international climate policy. A chapter on this topic is legitimate, as this role has, according to many observers, been substantive. For example, whereas environmental NGOs contributed to setting the issue of climate change on the political agenda, business NGOs have to some extent influenced the contents of the Kyoto Protocol mechanisms. To analyse such input and the impact of NGOs in more depth, this chapter makes a distinction between the *pre*-Kyoto and *post*-Kyoto eras (1990–1997 and 1998–2001, respectively). The reason for doing so is the fact that the Kyoto Protocol, although a first step, marks a turning point in international climate policy, not only in terms of goal setting, but also in terms of policy instruments and economic impact. Also, the role and effectiveness of environmental movements on the one hand and industry on the other have changed since 1997. Most remarkable is the more pro-active attitude of business towards international climate policy today compared to the early 1990s.

This chapter addresses the following research questions:

- What roles did NGOs, both environmental and business NGOs, play in international climate policy and how did these roles change over time (1990–2001)?
- What strategies did NGOs use to achieve their goals in the climate arena?
- To what extent were NGOs successful in influencing international climate policy?

In this chapter, we focus on the global level of NGOs, United Nations (UN) and intergovernmental negotiations. The chapter is organized as follows: first, in Section 2, below, we briefly discuss the concept of *NGO* and deal with the relationship between NGOs and the UN. Second, we examine the goals, strategies and effectiveness of environmental NGOs (ENGOs) and business NGOs (BNGOs) in the pre and post-Kyoto eras, taking into account the latest developments in international climate policy, such as COP6 bis in Bonn, July 2001 (Sections 3 and 4, respectively). The approach in these two sections differs, however. Whereas the section on the pre-Kyoto era is an ex post evaluation of NGO activities in the climate arena, consisting of an analysis of past NGO performance, the section on the post-Kyoto era comes closer to being an ex ante evaluation, with a formulation of possible strategies for NGOs in the near future. Section 5 concludes with a brief summary.

2   NGOS AND THE UN

The UN defines an international non-governmental organization (INGO) as: 'Any international organization which is not established by intergovernmental agreements' (Feld and Jordan, 1983, p. 227). This definition implies that a wide range of international organizations may be considered to be NGOs: churches, scientific unions, sports clubs, trade unions, political parties, the Scouts, business organizations, transnational corporations (TNCs), hybrid organizations of governments and private organizations (like the World Conservation Union/IUCN), environmental movements, peace organizations, liberation movements, terrorist groups and even the Mafia. Others conceptualize NGOs as pressure groups. According to Thompson-Feraru (1974, pp. 32–3): 'INGOs are international pressure groups ... that have the capacity or desire to influence the course of international relations'. This definition is more specific than the UN's since it covers the nature and goal of an NGO and excludes those organizations that do not generally wish to influence policies (such as sport clubs and the Scouts). Because this definition is more in line with the type of organization studied in this chapter, it is this latter definition that is used here.

NGOs and the United Nations have developed mutual relationships. Article 71 of the UN Charter states that the Social and Economic Council (ECOSOC) may consult international NGOs that are concerned with matters within the ECOSOC's competence, for UN conferences on special issues, among other things (Feld and Jordan, 1983). It has become standard practice for INGOs to participate in UN conferences and meetings, at least if they cover the issues on the agenda and are formally accredited by UN staff. Recently, this has come to include national and local NGOs as well (Willets, 1996). NGOs have observer status at such meetings, but they may generally make oral statements and disseminate written position papers. In addition, they may have access to working groups in which negotiations between countries take place, but this is decided on a case-by-case basis by the officials. Countries may include NGO representatives in their delegations at international meetings, but this is dependent on the willingness of individual countries and is not yet common practice. Apart from this formal access, NGOs can lobby in the corridors of such UN meetings, and non-accredited NGOs may organize protest activities around the buildings where the negotiations take place.

3   THE PRE-KYOTO ERA (1990–1997)

3.1   Environmental NGOs

According to Rahman and Roncerel (1994, p. 242), environmental NGOs contributed to the early climate debate in the 1980s in three ways: undertaking research, developing a database and pressing for ecological targets (explain). Organizations such as the World Resources Institute, Environmental Defense and the Stockholm Environment Institute conducted climate research themselves, while others, such as Greenpeace and Friends of the Earth, gathered and processed data to underpin the ENGOs' arguments. In March 1998, ENGOs established the Climate Action Network through which they started to co-operate and strategize. At the same time, many ENGOs pressured governments to take action. Examples of such action include the Toronto Conference in 1988 (a follow-up of the Brundtland report, where climate change drew a lot of attention), the Noordwijk Conference in the Netherlands in 1989 (on the effects of climate change on economic development and vice versa) and the Second World Climate Conference in 1989 in Geneva (where calls for a climate convention were launched). These meetings produced proposals for ecological targets of which the Toronto Target has become the most famous. It calls for industrialized countries to reduce their greenhouse gas emissions by 20 per cent in the year 2005, compared to 1988 levels. These conferences and targets were the predecessors of the formal negotiation of the Climate Convention, which started in 1991 (Bodansky, 1993). In addition, some countries, such as Austria and the Czech Republic, issued the Toronto Target as the keystone of their domestic policies on climate change. According to Paterson (1996) and others, this unilateral target setting, also inspired by domestic environmental pressure groups, functioned as a role model for other countries and as an international norm to guide global negotiations.

During the formal negotiations on the United Nations Framework

Convention on Climate Change (FCCC) and its early implementation in the period 1991–1997, NGOs were able to participate as observers, in both the sessions of the intergovernmental negotiation committees (INC) and the meetings of the conferences of parties (COPs). This participation was legitimized by Article 71 of the UN Charter as well as by UNCED decision 1/1, which was taken over by INC1 and which regulated NGO input. However, it was clearly stressed in the rules of procedure that NGOs had *no* negotiating role. Nonetheless, the actual procedures at the first negotiation session in Chantilly (INC1, 1991) were quite liberal and set a precedent for subsequent meetings. The Americans granted accreditation to all NGOs that wished to visit INC1. Consequently, more than a 100 were present there. The subsequent INCs were visited by approximately 50–75 NGOs, of which about one-third were environmental organizations. A large majority of these NGOs originated from the North, although the participation of Southern organizations increased over the years. Later on, COP 1 greeted 116 NGOs and COP 3, 236 (Oberthür and Ott, 1999). At COP 6 in 2000, the total number of NGO personnel was several thousand.

ENGOs tried to intervene in the negotiation and decision-making processes in the climate arena in order to affect policy outcomes. In so doing, they applied several strategies: lobbying, advocacy and protests being the most important ones (cf. Van Noort et al., 1987; Pleune, 1997). In this context, lobbying refers to informal and covert attempts to influence policymakers in the corridors of political events by transmitting, for example, information, knowledge, views, text proposals, sympathy or warnings. Advocacy refers to formally accepted and overt attempts to exercise influence in the political arena, again, by transmitting information, knowledge, views, text proposals, sympathy or warnings. Protest, finally, refers to informal and overt attempts outside, or at the margin of, political arenas to influence policymakers, for example by protest marches, sit-ins, consumer action, disturbances at meetings, and so on. During the INCs and COPs, ENGOs used all three strategies. For example, during COP 1 in Berlin (1995), Greenpeace occupied the huge chimney of a large German energy plant for several days, while colleagues were lobbying delegations and advocating their views in Berlin. At the same time, members of the youth organization A SEED organized a protest bike tour in the city, while others caused disturbances in some meetings in the conference with banners and verbal heckling. Most ENGOs, however, focused on advocacy and lobbying, co-ordinated by Climate Action Network (CAN). Applying these strategies in the period 1990–1997, ENGOs concentrated on specific issues in the climate negotiations, which they thought were the most important to defend or amend. Examples include the overall objective of the FCCC, its section on principles, the specific emissions targets of the FCCC, its implementation mechanisms and the need to strengthen the FCCC by means of a protocol after 1992 (Arts, 1998; Rahman and Roncerel, 1994).

The question at this point is whether the ENGO strategies have been effective with regard to these issues, in the sense that FCCC decisions changed in accordance with the NGOs' preferences. Some examples indeed support such political influence.[1] (An example is the issue of joint implementation.) The idea that industrialized countries can achieve emission reductions more efficiently through projects in the East and the South than at home, was launched at the Noordwijk Conference in 1989. Most ENGOs, however, immediately criticized this mechanism (Climate Action Network, 1992; Eco, 1993). They feared that the industrialized countries would buy off their domestic commitments and thereby avoid painful interventions at home by unloading the burden on the poor.

Although a few developing countries were in favour of the concept, others were confused about it. Various developing countries slowly adopted some of these critical NGO standpoints, due to the subtle diplomatic strategies of (mainly Southern) ENGOs. This impact of NGOs is also confirmed by several policy documents from the Government of The Netherlands (for example VROM, 1991, 1994), as well as by key negotiators at the time (including Vellinga, the Netherlands; Ripert, France; Luiz Meiro Filho, Brazil). Others, however, put this conclusion into perspective, and have emphasized, in turn, the influence of developing countries on ENGOs in the climate arena (Gupta, 1997).

Another example of ENGO influence is the emissions budget approach and international emissions trading. In 1996, the Dutch Government invited Environmental Defense to give a presentation to a UNFCCC workshop to present their views on an emissions budget and trading system for greenhouse gases, with multi-year targets for industrialized nations and project-based trading for developing nations. This presentation, given early 1996 in Geneva, launched the design of what ultimately became the Kyoto Protocol's emissions budget and the Kyoto mechanisms. BNGOs saw that such approaches as budgets and trading systems could work, but were concerned about the targets. Governments were interested, depending on the targets. Other ENGOs, such as the Deutsche Naturschutzring (DNR), and institutes, such as the London-based Royal Institute of International Affairs (RIIA), were positive about the concepts so long as the enabling mechanisms were simple and would be employed primarily by governments.

These two examples of political influence should, however, not close our eyes for other cases, in which ENGOs seemed to have failed to affect decision-making, such as regarding the principles, the general objective and the implementation mechanisms of the FCCC (Arts, 1998). Yet their intervening role was obvious in general, as was their political influence in some cases, in the pre-Kyoto era.

3.2 Business NGOs

ENGOs are generally more inclined to act as transnational actors in the international environmental arena than are corporations, as the case of climate change also shows (Levy and Egan, 1998). This is due to two factors: the inclination of corporations to focus (1) on the national level as far as environmental affairs is concerned and (2) on 'economic' regimes (WTO, MAI, NAFTA) rather than on environmental ones at the international level. In addition, corporations 'know' that their interests are generally better furthered by the political elite than those of other groups, because business is the driving force behind the economic growth on which the political system is so dependent. Yet business interfered in the UNCED process as well as in the climate negotiations in the early 1990s because corporations had learned that governments were sometimes more responsive to environmental demands than they would like (Chatterjee and Finger, 1994; Kolk, 1999).

Two examples of international pressure groups that appeared in the late 1980s are the World Business Council for Sustainable Development and the Global Climate Coalition (GCC), the latter being very visible in the climate arena from the early days. Founded in 1989, the GCC sought participation in the scientific and policy debate on climate change and was able to mobilize large segments of the coal, oil and car industries. It took the position that climate change was not yet proved, so (costly) measures were unnecessary. In other words, it advocated a 'business-as-usual' position. As a consequence, the GCC opposed a climate treaty in general – and certainly any binding international targets on the reduction of greenhouse gases (Kolk, 1999). Through the years, the GCC maintained its rather conservative viewpoint, which led to the departure of large multinationals, such as Shell, General Motors and Ford, from the coalition in the late 1990s (see Section 4).

To further its interests in the climate arena, the GCC hired professional lobbyists from Washington, DC, of whom David Pearlman became one of the most famous. His main strategy was to cling to the delegations from OPEC countries as much as possible, thereby exerting an influence on the climate talks. This strategy was effective in the sense that the OPEC countries became the major 'foot-draggers' in the negotiations. Together with the USA, they were not only able to water down the commitments in the treaty, but they also nearly killed the final and very subtle compromise text of the FCCC in the last session in May 1992, just before UNCED. The GCC's strategy was, on the other hand, ineffective in the sense that it alienated the GCC from the other negotiators. As one commented in an interview with one of the authors: 'Sometimes, BNGOs crossed the boundaries of what governments consider proper behaviour. They tried to push it too hard. They got too far involved in the governmental process.' This quote refers to some negotiation sessions, in which Pearlman actually sat at one of the OPEC delegation tables, and instructed the delegates what to say and what to vote for. This extremely annoyed other delegations, all the more so since the message was so negative and obstructed progress in the negotiation process.

Another example of the GCC's ineffectiveness was the adoption of Article 4 of the FCCC (Arts, 1998). This article includes a diluted version of the stabilization target, which says that industrialized countries should stabilize their greenhouse gas emissions in the year 2000, compared to 1990 levels. In the FCCC, however, this target is written in very ambiguous legal language as a compromise between the USA, Europe and Japan. To the propagandists of climate change policy – for example ENGOs and European countries – this outcome was a shock; they had hoped that stronger targets would be included in the FCCC. But for business it was a shock as well, because they had anticipated an outcome without any target, given the positions of the USA and OPEC. Hence, BNGOs failed to achieve all of their goals in this case.

Although the business lobby in the early days of the climate negotiations was mainly determined by the GCC, the picture provided above is not complete. Some parts of industry, such as the 'green-energy technology' sector and the insurance industry, were very supportive of climate measures right from the start (Kolk, 1999). The insurance industry's position can be understood because changes in climate change – possibly leading to floods, hurricanes, droughts, crop failures, and so on – pose enormous potential for large-scale damage to property and to subsequent claims on the insurance industry. To a large extent, this sector co-operated within the financial institutions initiative that was established in 1992 under the auspices of UNEP. In Berlin, at COP 1, the sector even co-operated with Greenpeace in launching a joint campaign to support substantive measures. This cooperation with an ENGO proves that business interests had to be more broadly defined than those endorsed by the GCC.

BNGOs became ever more engaged in the process of the Ad Hoc Group on the Berlin Mandate (AGBM), preparing for the Kyoto Protocol in the period 1995–1997. In particular, through the US delegation, they made their preferences very clear for comprehensive economic studies and analysis of potential policies, for moderate reduction norms and for market-oriented instruments to achieve these norms (for example emission trading). Although it is difficult to assess the exact impact of business on these negotiations, the contents of the Kyoto Protocol, with its emphasis on flexibility, seem to confirm hypotheses about the influence of corporate power in the climate negotiations, although business was again – just as in the case of the FCCC itself – surprised that the binding norms went as far as they did (Balanya et al., 2000; Levy and Egan, 1998).

4   THE POST-KYOTO ERA (1997–2001)

4.1   Validation of the Kyoto Agreement

Kyoto resulted in absolute, binding emission commitments significantly below the 'business-as-usual' trends of many industrialized countries. Most ENGOs did not expect that to happen. But they were still not satisfied because the emissions cuts (–5 per cent in total) were far from what is needed to tackle climate change. Because of the pressure they mobilized during the negotiations and decision-making process and the political momentum that evolved in Kyoto, the ENGO community can to a large extent be credited for the fact that Kyoto resulted in a Protocol. On the other hand, the Kyoto deal was criticized by most ENGOs because of so much potential flexibility and so many 'loopholes' (Greenpeace International, 1997). The ENGOs noted that the use of the Kyoto mechanisms, the sequestration of CO_2 (sinks) and the sale of surplus greenhouse gas emissions by Russia and Ukraine (called 'hot air' by the ENGOs) could lead to marginalization and postponement of actions by industrialized countries.

In their analysis, these ENGOs used the work of such institutes as the London-based Royal Institute of International Affairs (RIIA) and the Dutch Rijksinstituut voor Volksgezondheid en Milieuhygiene (RIVM), which sketched the worst-case scenarios based on the assumption that all parties would make full use of sinks and Kyoto mechanisms. The Protocol could further help the survival of nuclear generation; in particular, it could help finance nuclear energy by using it as a clean development mechanism. Some ENGOs, however, also noted the need for market mechanisms to achieve timely and high reductions, as well as the need to include forest activities. They also welcomed the appearance of the flexible mechanisms and sinks in the Protocol text. Although they stressed the notion of restriction to omit levels above national caps, they felt that flexibility should be guaranteed within these boundaries. There was hardly any criticism on the 'EU bubble', a provision that gives the EU the opportunity to distribute the common EU emissions commitment over its member states (allowing southern member states significant increases in emissions) while internal compliance rules still need to be set.

BNGOs were surprised, too. They did not envisage an outcome from Kyoto with absolute, quantified commitments for greenhouse gas reductions, which implied for nearly all industrialized countries the need to curb 'business-as-usual' emissions and to attack the fossil-fuel-based economy. British Petroleum was an exception to this, applauding the Kyoto outcome and putting its own Kyoto target forward. But in general, companies – in particular, those from the US – felt that the Kyoto Protocol was too ambitious.

On the other hand, ENGOs felt that the Protocol was too flexibile. This shows that the Kyoto deal was 'very much an agreement struck by governments, negotiating for what they felt to be possible and appropriate. The future of international co-operation on climate change depends on whether and when the Protocol will be ratified and implemented by the major players' (Grubb et al., 1999).

Against this background, it is interesting to analyse how ENGOs and BNGOs develop their strategies for the next phase of climate policymaking: elaborating the provisions of the Protocol and preparing for its ratification and implementation. As Grubb et al. (1999) note, implementation will definitely depend on the BNGO and ENGO communities. A parallel can be drawn with regard to the North American Free Trade Agreement (NAFTA), where the environmental community deserved substantial credit for bringing environmental safeguards and mechanisms into the package (although one can argue that these safeguards are not sufficient). Now ENGOs play a key role in determining the success of its implementation (Ferber et al., 1995). Here, one important precedent is the recent ruling of a NAFTA panel that NGOs can, in principle, have intervener status in investor-state cases. In the case of *Methanex v. California* under Chapter 11 of NAFTA, the International Institute for Sustainable Development and Earth Justice filed petitions[2] (International Environment Reporter, 2001).

Environmental NGOs
Whereas the focus of ENGOs up to and in Kyoto was on the quantified targets for greenhouse gas emissions, their focus after Kyoto has substantially changed. Originally, the general strategy was to advocate limits on the use of the Kyoto mechanisms, on the role of forests and on selling 'hot air', as well as to advocate preferences on reduction projects, on implementing policies, on renewables and on excluding nuclear energy. Sometimes this resulted in supporting EU views and bashing those of the US. The positions of the larger international ENGOs – Greenpeace International, World Wildlife Fund

(WWF) and Friends of the Earth – have hardened recently, given their statement 'make or break the Kyoto Protocol'. Their representatives have announced that they will leave the Kyoto Process if governments do not meet their demands at COP 6 (November 2000), referring specifically to the use of nuclear energy and sinks. On the other hand, ENGOs have focused pragmatically on getting the Kyoto Protocol ratifiable, particularly at COP 6 bis in Bonn (July 2001), while keeping it environmentally credible and enforceable.

Climate Action Network (CAN) embraces a diversity of interests and positions, but the dominant feeling among the mainstream ENGOs is a dislike of the Kyoto mechanisms and forest-carbon offsets. The Kyoto mechanisms have been prominently portrayed in various CAN statements and individual papers as selling off and claiming unwanted ownership over clean air. In addition, forest projects have been portrayed as an allowance to increase emissions. Proponents of the Kyoto mechanisms in the ENGO community – some based in Latin America and some in the US, not so many in Europe – were met with scepticism. In reacting this way, ENGOs have neglected the opportunities offered under the Kyoto mechanisms and have ignored the potential positive effects of the mechanisms and forest activities, especially since much emissions cuts are needed in the second commitment period.

With a sound national registry/monitoring system and an effective enforcement mechanism with adequate incentives, the Kyoto mechanisms could ensure compliance and lead to innovations and reductions with a lower price for all. Nentjes and Boom (2000) suggest that ENGOs should also focus on the design of emissions trading schemes. But ENGOs have not found it very easy to become effectively engaged in all the aspects of the climate issue as it has become steadily more complex. There is a mixed approach towards the Kyoto mechanisms and the role of sinks and forests. ENGOs have remained focused on issues of environmental integrity, targets and equity but not on implementation. These problems are, as noted by Grubb et al. (1999), embedded in the Protocol under the name of flexibility.

One may question whether this defensive attitude is the most effective one, and whether it is correct, considering the current background. The strategy of the NGOs could be more effective if it were adapted to reflect current developments in the climate arena. First of all, the implementation of the Protocol requires ratification by nearly all major industrialized countries. The current resistance to ratification by the Bush administration in the US might still lead to resistance from Canada, Japan and the Russian Federation. Despite the 'success' in Bonn, the result could be an inert treaty. ENGOs have a stake in making the Protocol ratifiable and an important role.

When international discussions do not lead to solutions, the greatest hurdle – that the Protocol will not be implemented because of an insufficient number of ratifying parties – becomes the major focus of attention. This seems to have happened among the ENGOs in Bonn, where they felt that the process itself was at stake.

Second, as the Kyoto Protocol moves into another policy phase and heads towards implementation,[3] the practical design of mechanisms and measures are higher on the agenda. After the closure of the Bonn Summit, Greenpeace International targeted their actions on implementation, with the slogan 'You might have loopholes, but you don't use them'. In addition, ENGOs have many good ideas about implementation. In a recent study, the need for ENGOs to revisit their role is stressed (Cozijnsen, 2000a). ENGOs could enhance their effectiveness by working on fact-finding, disclosure and monitoring, as well as by working with companies. After all, publications and actions of this kind by ENGOs have an enormous impact, as shown by the recent WWF study on the status of the existing EU plans for climate policy and the studies released before the Bonn Summit on the economic benefits for the EU, Japan and the US if they were to implement the Kyoto Protocol (see www.panda.org). Besides stressing environmental integrity and credibility regarding the market mechanisms in the Protocol, ENGOs could focus on sound and prompt implementation of the Protocol's commitments, all the more so since this could even lead to over-compliance (that is, reducing emissions more than laid down in the commitments). This would be even better for the environment. Such over-compliance has already been witnessed in the prototype SO_2 Allowance Programme in the US.

Third, regarding climate change and global environmental issues in general, individual countries have less room to manoeuvre because of the globalization of markets. In Europe, more decision making is done in Brussels, more firms operate in an international market and corporate acquisitions show a growing number of transnational companies. In addition, electricity and gas markets are being restructured and liberalized in the US, the EU and economies in transition. This implies that market parties will have to be more responsible for, and more involved in, international environmental policy, as well as there being a growing importance of market-based instruments in policymaking. This may stimulate ENGOs to enter into market-related activities. NGOs could, for instance, play a monitoring or auditing role (see the work of Environmental Resources Trust, www.ERT.net). Some ENGOs are involved in the verification and accreditation process of projects. CAN and the World Resources Institute issued proposals for public participation in the clean development mechanism. WWF already goes one step further by promoting green-label projects for fast-track approval of clean development mechanisms (CDMs) (Salter, 2001). Other examples where ENGOs play a market role include the following: (1) WWF's work on green electricity labelling, (2) Greenpeace International's 'greenfreeze', 'green energy' and 'smart-car', and (3) The Green Power Market Development Group, a partnership between WRI, Business for Sustainable Responsibility and 11 major US companies.

Fourth, this 'marketization' is enforced by another development – the withdrawal of government involvement in various areas (Cozijnsen, 2000b), especially in European countries. The US government was already less involved in environmental issues, although this situation has changed since the US Senate has blocked environmental initiatives that the administration has pushed. Because of this, representatives of Greenpeace US and Environmental Defense recently agreed that working with the corporate world, rather than in the political process, is more effective (Allen, 2000). This all means that governments have to focus increasingly on the core tasks of setting targets and organizing enforcement, while implementation will rely more on the participation of stakeholders and the use of market-based instruments. Examples relevant to climate change policy are, according to one study, CO_2 taxation, green certificates, product demands and emission trading as provided for in the Kyoto Protocol (Cozijnsen, 2000b). This study also notes a potential increase in the effectiveness of ENGOs if they work with companies, provided they stay close to the Kyoto Process itself. This means that actions must contribute to meeting Kyoto targets. A good example of this is WWF's Climate Savers Programme (http://www.panda.org/climate/savers.cfm) and ED[F]'s Partnership for Climate Action (with BP, Shell, Pechiney, Dupont and others; see http://www.environmentaldefense.org/PCA/). Under conditions of maintaining their credibility and financial independence, ENGOs can indeed encourage companies to work on baselines and reductions (Fastiggi, 1999). Until recently, partnerships with companies were not linked with absolute targets, such as the Kyoto commitments. Therefore the challenge is to show that ENGOs can work with individual companies to meet such targets. At the same time, ENGOs should remain keen on their position *vis-à-vis* industry, given the fact that their mutual power relationships are quite uneven – co-optation of environmental organizations by firms remains a threat for 'green alliances' (Arts, forthcoming).

Business NGOs

The situation has also changed for industry since Kyoto. The corporate world has changed: insurers are worried about their costs rising because of climate change, gas companies see opportunities, electric utility and car companies envisage transitions ahead, and oil and coal production are increasingly vulnerable. Producers of fuel cells see two triggers for their deployment: lower greenhouse gas emissions and less dependency on fossil fuels. Some of these companies have taken the threat of climate change to heart and have committed themselves to reducing emissions (BP, Shell, other Pew Climate Members, IPIECA) and to enhancing the solar market share. US–EU mergers (BP with Amoco, Daimler with Chrysler) and cross-border energy acquisitions play a role too. In general, European companies are not against the Kyoto Protocol, as long as competitiveness is not affected (compromised). Auditing companies are preparing to verify emission baselines and reductions; consultancies and brokers specializing in carbon offsets in the US, Canada, Australia and Europe are creating carbon funds; banks and insurers are starting carbon funds; and energy companies are bringing carbon-free energy to consumers (Cozijnsen, 2001).

Some pro-active groups formed by the private sector have a stake in clean energy, co-generation and efficiency, examples of which are the US Business Council on a Sustainable Energy Future, and the European Business Council on Sustainable Energy/E5. They see opportunities created by the ambitious targets of the Kyoto Protocol. What's more, they would even favour limits on the use of the Kyoto mechanisms, accept liability on the side of buyers of units, and reject trade-offs through sinks (Kolk, 1999). They would also accept taxation on CO_2. They want to give an example to the corporate world, ENGOs and governments, as their E-55 initiative proved in Bonn. More that 55 companies pushed the parties to ratify the Protocol. One of their members even delivered carbon-free electricity to the Conference Centre in Bonn.

A second – also pro-active – group is composed of large fossil fuel companies and big energy users, who are major contributors to the climate problem, but who are also aware of the issue and are prepared to reduce emissions. These companies want to set the tone rather than wait for governments to allocate commitments and to come up with answers. These companies act through organizations such as the Pew Center on Climate Change to raise awareness, to analyse responses, to take unilateral commitments to reducing emissions (examples of this include BP, Shell, Dupont), to start emission trading schemes at the company level (BP, Shell), or to join other organizations in developing emission inventories and monitoring protocols for companies (WRI/WBCSD).

The third group – still defensive in nature – consists of the rest of the coal and fossil fuel industry. These companies claim that their economic position is at stake, focus on maximum flexibility, do not accept any limits on the use of the Kyoto mechanisms (which would result in the lowest price for global greenhouse gas reduction and prevent high transaction costs) and would like to minimize government and administrative involvement. They also support self-reporting, voluntary agreements and liability on the part of the sellers of units. This part of the business community is not helpful to the proactive part, since their claims could raise concerns about the credibility of a company's use of the mechanisms. ENGOs argue that BNGOs can already enjoy flexibility through the Kyoto mechanisms, because overall reductions would be cheaper if all companies reduced emissions at their own plants. According to ENGOs, companies should not try to get rid of all transaction costs. Some ENGOs even prefer to put levies on the use of emissions trading, joint implementation and CDM projects in order to put pressure on domestic emissions reductions.

Recent history shows that the first two groups of BNGO actors could be organized and mobilized, and could create a common interest. They want to set the tone for the implementation of the Kyoto Protocol in the future in a way that fits their core business. The third group has been less effective within the development of climate policy because of the lack of pro-active common interests. But, led by EXXON US, they succeeded in moving the Bush administration away from the Kyoto Process. Economic concerns were taken over by President Bush when he said that he felt the Protocol was 'fatally flawed'. Further proof is the recent rapid shift of the Global Climate Coalition (profound nay-sayers) from a lobby to a knowledge forum, moving away from involvement in the process. In general, one can conclude that BNGOs have been more effective in setting the agenda for the negotiations after Kyoto because greater priority has been given to the Kyoto mechanisms than to the adequacy of commitments and future targets. Nevertheless, because of the influence of ENGOs, environmental integrity was has continued to be an issue when the Kyoto mechanisms have been discussed.

A recent study compared the awareness and strategies of companies in three sectors in the US and Europe (Van der Woerd et al., 2000). It focussed on triggers within the business community that could change behaviour: opportunities for individual companies, compatibility of the Kyoto mechanisms, accessibility to new technologies and exploitation of non-CO_2 gasses. Although not included in the study's conclusions, it looked for co-operative programmes between business and ENGOs. In particular, it appeared that pioneers among companies have such programmes. This is true in the oil industry. In the automobile industry, some European companies have partnerships, but in the US there is only one. In the chemical industry, most US and European companies have advisory contacts with ENGOs. In the bank and insurance sector, four out of five European companies have partnerships, whereas none of the five responding US companies reported such arrangements. Against the background of the current position of the Bush administration, this study shows that some companies regret the US stance because they miss opportunities, while others feel comfortable with the lack of urgency.

The issue of corporate responsibility can be regarded as the wider picture. This issue came up in the preparation of UNCED, and it will continue to be an issue for several years (SER, 2001). With regard to transnational corporations (TNCs) in particular, there are a number of issues that will have to be addressed: acceptable corporate behaviour, standards for disclosure, contribution to climate actions, and so on. We realize that in the ongoing international negotiations, the discussions are mainly about the contribution and role of governments, business entities and civil society. So far, there is no link with what would be expected from transnational companies. Shell and BP introduced emissions trading amongst business units as an internal instrument to meet their own reduction targets, hoping that an international emissions trade regime would be compatible with their own schemes. But practice shows that these businesses will continue to face different environmental rules and targets in the countries where their headquarters are located. For example, BP is still unclear how the participation of their UK assets in the UK emission-trading scheme will interact with their global, internal arrangement for emission trading. BP expects that everything will depend on the ultimate price of carbon reduction. In this instance, TNCs will contribute more to climate change policies using transnational market mechanisms, than they will when falling under different domestic policies. They will also have to lobby for each national government to implement comparable schemes.

Apart from climate awareness and commitments to reduction, *per se*, BNGOs can choose to be pro-active on climate change because of indirect advantages, such as gaining a better stock price in the green business market. Trouw (a Dutch daily newspaper) and the ASN Bank maintain indexes of the green business market, as does the Dow Jones Sustainability Group. A more recent study compares the revenues of large Dutch companies with the costs they have when limiting their emissions (www.tme.nl) Furthermore, companies in general can gain environmental credibility and increase their attractiveness for investors, young executives and customers.

5   CONCLUSION

NGOs, both environmental and business, have played major roles in the development of international climate policy since the 1980s. They have raised awareness among their constituencies, made science more understandable, advocated their views among the public and politicians, lobbied policymakers, organized protest marches, influenced the contents of policy documents, been involved in project implementation and experimented with the Kyoto mechanisms. In so doing, ENGOs originally tried to 'curb the trends' of global warming through strengthening measures aimed at controlling climate change as much as possible, while BNGOs advocated 'business as usual' by blocking or softening these measures. Since 1995, this situation has, to some extent, changed. Business has become much more pro-active and supportive in regard to climate measures. They now accept climate change as a problem, and they see chances to combine emission reductions with market opportunities. Environmental organizations, on the other hand, have become more responsive to flexibility, differentiation, and market solutions and are seeking ways to shape their role in a more market-oriented approach. This has made co-operation between business and ENGOs possible. At the same time, it remains, in our opinion, of the utmost importance for ENGOs to retain their critical 'watchdog' role in international climate policy, while co-operating with governments and business.

NOTES

1. One should keep in mind that political influence is difficult to assess and always disputed in political science. The examples in this chapter are taken from Arts (1998). In determining the extent of ENGO influence in the climate arena, he applied methodologies that are broadly accepted in the political science community (see also Arts and Verschuren, 1999). Yet he adds that conclusions on political influence always remain 'educated guesses'.
2. Methanex is seeking compensation for the California government's 1999 ban on the use of gasoline additive methyl tertiary butyl ether. Two environmental organizations filed a petition seeking amicus status to present written submissions, to see the briefs and counter-briefs filed by the parties, to participate in oral arguments, and to sit as an observer during oral arguments presented by the parties.
3. This does not imply that the formulation of policies is over; on the contrary, mechanisms should be implemented and projects should be put into operation now, while policies are still developing.

REFERENCES

Allen, A. (2000), 'Gore or Bush? Who cares? Not environmentalists', *Salon.com*, October 23, 2000, see: http://www.salon.com/health/feauture/2000/10/23/globalwarming/index.html.

Arts, B. (1998), *The Political Influence of Global NGOs: Case Studies on the Climate and Biodiversity Conventions,* Utrecht: International Books.

Arts, B. (forthcoming, 2002), 'Green Alliances of Business and NGOs: New Styles of Self-regulation or Dead-end Roads?', *Corporate Social Responsibility and Environmental Management* (accepted for publication).

Arts, B. and P. Verschuren (1999), 'Assessing Political Influence in Complex Decision-making: An Instrument Based on Triangulation', *International Political Science Review*, **20**(4), 411–24.

Balanya, B., A. Doherty, O. Hoedeman, A. Ma'anit and E. Wesselius (2000), 'The Weather Gods: Corporations Profit from Climate Change', in B. Balanya, A. Doherty, O. Hoedeman, A. Ma'anit and E. Wesselius, *Europe Inc.: Regional and Global Restructuring and the Rise of Corporate Power*, London: Pluto Press.

Bodansky, D. (1993), 'The United Nations Framework Convention on Climate Change: A commentary', *Yale Journal of International Law*, **18**, 453–559.

Chatterjee, P. and M. Finger (1994), *The Earth Brokers: Power, Politics and World Development*, London: Routledge.

Cozijnsen, J. (2000a), 'Energy, Accession and Climate Change', Paper prepared within the joint WRI/REC project 'Capacity for Climate Change'.

Cozijnsen, J. (2000b), *Derden en nieuwe instrumenten (The Third Party Position in a Changing Environmental Permit System)*, The Hague: Ministry for Public Housing, Spatial Planning and the Environment (VROM).

Cozijnsen, J. (2001), 'The Development of Post-Kyoto Emissions Trading Schemes in Europe: An Analysis in the Context of the Kyoto Process', in *Greenhouse Gas Market Perspectives*, New York and Geneva: UNCTAD (see also htto://www.unctad.org/ghg/index.html).

Fastiggi, E. (1999), *Catalyzing Environmental Results: Lessons in Advocacy Organization-business Partnerships*, Washington: Alliance for Environmental Innovation.

Feld, W. and R. Jordan (1983), *International Organizations: A Comparative Approach*, New York: Praeger.

Ferber et al. (1995), 'Building an Environmental Protection Framework for

North America: The Role of the Non-governmental Community', in *Green Globe Yearbook 1995*, Oxford: Oxford University Press.

Greenpeace International (1997), Press release, Amsterdam: Greenpeace International, 11 December, 1997.

Grubb et al. (1999), *The Kyoto Protocol: A Guide and Assessment*, London: Royal Institute of International Affairs.

Gupta, J. (1997), *The Climate Convention and Developing Countries: From Conflict to Consensus?*, Dordrecht: Kluwer.

International Environment Reporter (2001), 'NAFTA Panel Says NGOS Can Intervene In Cases Brought for Arbitration Purposes', in *International Environmental Reporter*, **24**(3), 31 January 2001, Washington: The Bureau of National Affairs, Inc.

IPIECA (1991), *Global Climate Change: A Petroleum Industry Perspective*, London: International Petroleum Industry Environmental Conservation Association.

Kolk, A. (1999), 'Multinationale ondernemingen en internationaal klimaatbeleid (Transnational Corporations and Climate-change Policy)', *Milieu*, **1999**(4), 181–91.

Levy, D.L. and D. Egan (1998), 'Capital Contest: National and Transnational Channels of Corporate Influence on the Climate Negotiations', *Politics & Society*, **26**, 337–61.

Nentjes, A. and T.J. Boom (2000), 'Emissions Trading with a Cap and Without a Cap.' *Change*, **53**, 4–7, August–September 2000.

Oberthür, S. and H.E. Ott (1999), *The Kyoto Protocol: International Climate Policy for the 21st Century*, New York etc.: Springer-Verlag.

Paterson, M. (1996), 'IR Theory: Neorealism, Neoinstitutionalism and the Climate Change Convention', in J. Vogler and M. Imber (eds), *The Environment and International Relations*, London: Routledge, pp. 59–76.

Pleune, R. (1997), *Strategies of Dutch Environmental Organizations: Ozone Depletion, Acidification and Climate Change*, Utrecht: International Books.

Rahman, A. and A. Roncerel (1994), 'View from the Ground Up', in I.M. Mintzer and J.A. Leonard (eds), *Negotiating Climate Change: The Inside Story of the Rio Convention*, Cambridge: Cambridge University Press.

Salter L. (2001), *Carbon Labelling and JI/CDM*, presentation on behalf of WWF at the Workshop on CDM/JI, organised by the European Commission, Brussels, 22/23 March, 2001.

SER (2001), *Corporate Social Responsibility: A Dutch Approach*, Assen: Van Gorcum.

Thompson-Feraru, A. (1974), 'Transnational Political Interests and the Global Environment', *International Organization*, **28**, 31–60.

Van der Woerd, F., C. de Wit, A. Kolk, D. Levy, P. Vellinga and E. Behlyarova (2000), 'Diverging Business Strategies Towards Climate Change: A USA-Europe Comparison for Four Sectors of Industry', *Global Change Report 410200052*, Bilthoven: Dutch National Research Programme on Global Air Pollution and Climate Change.

Van Noort, L. Huberts and L. Rademakers (1987), *Protest en pressie. Een systematische analyse van collectieve actie (Protest and Pressure. A Systemic Analysis of Collective Action)*, Assen: Van Gorcum.

VROM (1991), *Report of the Dutch Delegation to INC3*, The Hague: Ministry for Public Housing, Spatial Planning and the Environment (VROM).

VROM (1994), *Report of the Dutch Delegation to INC10*, The Hague: Ministry for Public Housing, Spatial Planning and the Environment (VROM).

Willets, P. (1996), 'From Stockholm to Rio and Beyond: The Impact of the Environmental Movement on the United Nations Consultative Arrangements for NGOs', *Review of International Studies*, 22, 57–80.

13. Climate options for the long term: possible strategies

Bert Metz, Arthur Mol, Magnus Andersson, Marcel M. Berk, Jelle G. van Minnen and Willemijn Tuinstra

1 INTRODUCTION

The United Nations Framework Convention on Climate Change (UNFCCC), ratified by 186 nations and in effect since 1994, states that its objective is to stabilize greenhouse gas concentrations in the atmosphere at such a level and within such a timeframe that no dangerous interference with the climate system occurs that would threaten food supply, natural ecosystems and sustainable development (UNFCCC, 1992). There is a scientific consensus that for such stabilization, global greenhouse gas emissions have to decline to less than 50 per cent of current values (IPCC, 1995), while projections based on plausible developments point to significant increases of up to about 400 per cent over the next 100 years (Nakicenovic et al., 2000). Thus the challenge of controlling the risks of climate change in accordance with international agreements means that drastic emission reductions are inevitable and that they may have to be realized over a relatively short period of time. Is this possible? And what would it require? What does it mean for developing countries trying to combat poverty and improve the living conditions of their people? And what does it mean for industrialized countries that have built their current prosperity on fossil fuels and energy-intensive development patterns? Those are the key questions when looking at ways to control climate change in the long term.

The COOL (Climate OptiOns for the Long term) project aimed to answer these questions by identifying (elements of) viable strategies for drastic (that is 50–80 per cent by 2050) greenhouse gas emissions reductions for the Netherlands, in a European and global context. A stabilization level of 450 ppm CO_2 (or roughly 550 ppm of CO_2-equivalent, including the effect of other greenhouse gases) would require a 15–25 per cent reduction of global emissions by 2050 compared to current values. And it would require a much higher (50–80 per cent) cut in greenhouse gas emissions in industrialized countries (including Europe), given the much higher emissions in per capita terms of industrialized countries and the need to create room for developing countries to improve their welfare significantly. This chapter covers the global and European aspects of the COOL project. For the findings on the sectoral level in the Netherlands, see Hisschemoller and van der Kerkhof (2001).

In developing pathways for a future with low greenhouse gas emissions and assessing the various options that can contribute to such a path the COOL project deviated from more conventional studies on two major methodological points. First, in assessing the various pathways and options it used a participatory approach, often labelled participatory integrated assessment (PIA). Second, in designing and constructing the pathways and options the COOL project adopted a backcasting rather than a scenario or forecasting methodology.

1.1   Participatory Integrated Assessment

Participatory Integrated Assessment starts from the idea that designing policies and strategies and assessing their contribution to reaching ultimate goals improve if these activities and processes are not restricted to a small inner circle of scientists and policymakers, but also involve interest groups and civil society. Especially with respect to complex, unstructured environmental problems, where scientific investigations have their limitations, a broader involvement of different actors would add meaningful information, new insights and a less narrow and less technocratic view on causes and solutions. More specifically, the arguments in favour of a more participatory approach can be summarized as follows (cf. Funtowics and Ravets, 1993; Dunn, 1994; Fisher, 1990; Hisschemöller and Hoppe, 1996; Wynne, 1996; Irwin, 1995; Dryzek, 1987; Mayer, 1997):

- it helps to bridge the gap between a scientifically-defined environmental problem and the experiences, values and practices of actors who are at the root of both cause and solution of such problems;
- participation helps in clarifying different, often opposite, views and interests regarding a problem, making problem definitions more adequate and broadly supported;
- participation has an important learning component for the participants;
- participation in the scientific assessment may improve the quality of decision making, not by taking over the role of scientific expertise but by adding to it and supplementing it with other dimensions. As such it

increases feasibility, prevents implementation problems, establishes commitment among stakeholders, increases the democratic content, and so on.

At the same time, participation can also disturb policymaking and assessments, as several scholars point out (Seley, 1983; Douglas and Wildavsky, 1983; Van Thijn, 1997): stakeholders may be considered incompetent; they may have a major interest in a specific decision or non-decision and thus frustrate optimal solutions; participation can result in decisions that favour the more powerful and resourceful groups; and the process of participatory policy-making or assessments may become unproductive when it is time-consuming, conflict-enhancing and cannot live up to the promises of democratic decision-making that it implies.

1.2 Backcasting

The second main methodological characteristic of the COOL project is a backcasting approach. Backcasting was initially developed primarily in energy studies, being extended later to other sectors such as transport and infrastructure (Drehborg, 1996). Rather than analysing possible scenarios for the long term and in that way forecasting the possible and likely development paths of a social system, backcasting takes a desirable long-term future and analyses backwards to identify what different actions, steps and conditions are crucial at various points in time to reach that final normative objective. Backcasting should thus be seen in the context of Constructive Technology Assessment (Schot, 1992), Transition Management (Rotmans et al., 2000) and more normative branches of Ecological Modernization (see Mol and Spaargaren, 2000).

Within the COOL project the methodology of backcasting has been operationalized in five main steps. The first step involves the definition of the problem and setting the criteria for a solution (climate change caused by greenhouse gas emissions and 80 per cent CO_2 emission reduction between 1990 and 2050 for the OECD countries, respectively). The second step is the development of a so-called 'image of the future', an image of the social system or sector in 2050 that meets the requirements set by the criteria. The third step is the path analysis. An analysis is made of the pathway from the image of 2050 back to today to identify the transformations that are necessary, the lead-time for different options that can contribute to such transformations (for example rate of development and diffusion of fuel-cell technology), the crucial actors and conditions that make such options work, and the time to start to make these options contribute to the final image. Step four is a comparison of current trends – not only in greenhouse gas emissions, but also in energy production and use, transport demand and supply, agricultural production and consumption – and the desirable trends according to the path analysis. This so-called gap analysis provides us with ideas about the necessary policies for the coming years. The final step involves the formulation of an integrated strategic vision, in which the outcomes of the previous four steps are integrated into a single document that brings together the possible options and measures that need to be taken, the time paths for these options and measures, the conditions that support these options and measures and the coalition of actors that is crucial for implementing these options and measures.

Backcasting is a tool that informs us of the possibilities for radical change and highlights the pros and cons, the conditions for and consequences of different strategies, options and solutions that contribute to these radical changes.

Section 2, below provides a brief overview of the relevant literature. Section 3 describes the outcome of the COOL global dialogue in terms of key strategic findings for achieving drastic emission reductions and the implications for the short term. Section 4 does the same for the COOL Europe dialogue. Section 5 summarizes the main lessons on the general substance of the exercise and on the usefulness of the dialogue and backcasting methodologies.

2   WHAT DO WE ALREADY KNOW ABOUT LONG-TERM STRATEGIES FOR DRASTIC EMISSIONS REDUCTIONS?

The literature on quantitative studies into long-term strategies to control climate change is still fairly limited. The latest assessment by IPCC (Morita et al., 2001) provides a useful overview. One of the most important issues is the background against which long-term studies are performed. Until recently, most studies assumed a kind of 'business as usual' baseline as their reference point, leading to a narrow range of possible developments. However, the recent IPCC Emission Scenarios (Nakicenovic et al., 2000) accept that over time the world may develop in very different directions, leading to societies with very different characteristics and different distributions of welfare between countries. This is important because there is a growing recognition that the socio-economic conditions in countries determine their 'mitigative capacity' (Yohe, 2001); that is the degree and type of emission reduction effort depend on the socio-economic situation.

The IPCC scenarios without climate change mitigation show a large variation in CO_2 profiles over the next 100 years depending on economic and demographic developments and choices in the energy system (see Figure 13.1). The 450 ppmv stabilization case requires drastic and early emission reduction with very rapid emission reduction over the next 20 to 30 years. The higher the level of the baseline emissions, the larger the divergence from the baseline that is needed and the sooner it must occur. In other words, the world we live in will have a major impact on the possibility to control climate change, regardless of the mitigation action itself.

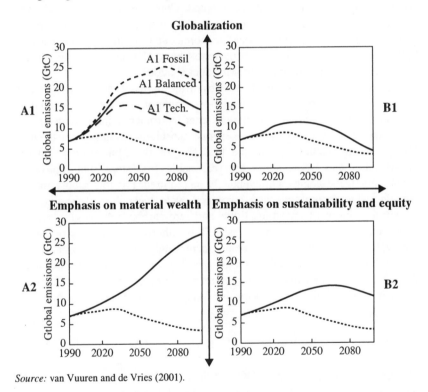

Source: van Vuuren and de Vries (2001).

Figure 13.1   Global CO_2 emission trajectories for the different IPCC-SRES marker scenarios compared to a 450 stabilization emission profile (dotted line)

An important policy question concerns the participation of developing countries in emission mitigation. Studies suggest that the stabilization target and the baseline emission level are both important determinants of the point in time when developing countries emissions might need to diverge from their baseline. For a 450 ppm CO_2 stabilization level industrialized countries alone cannot keep global emissions low enough. The extent to which developing countries would already have to be below baselines by around 2050 depends greatly on how the North and South share the effort and on the world we live in (Berk et al., 2001).

No single measure will be sufficient for the timely development, adoption and diffusion of mitigation options to stabilize atmospheric greenhouse gases. Instead, a portfolio based on technological change, economic incentives and institutional frameworks should be adopted. Combined use of a broad array of known technological options has a long-term potential which, in combination with associated socio-economic and institutional changes, is sufficient to achieve stabilization of atmospheric CO_2 concentrations in the range of 450–550 ppmv or below (Morita et al., 2001).

Mitigation options assumed by studies differ, in many cases also driven by the quantitative models used. However, common features of mitigation strategies include large and continuous energy-efficiency improvements and afforestation as well as low-carbon energy, especially biomass, over the next 100 years and natural gas in the first half of the 21st century. Energy conservation and reforestation are reasonable first steps, but innovative supply-side technologies will eventually be required. Possible robust options include using natural gas and combined-cycle technology to bridge the transition to more advanced fossil-fuel and zero-carbon technologies, such as hydrogen fuel cells. Solar energy as well as either nuclear energy or carbon removal and storage would become increasingly important for a higher emission world or lower stabilization target. It is important to note that under the more environmentally-friendly reference scenarios significant changes in the energy system are already occurring independently of mitigation policies.

3   ELEMENTS OF LONG-TERM GLOBAL STRATEGY

The global dialogue of the COOL project involved a group of about 25 representatives of stakeholder groups in the global climate change debate: governments in industrialized and developing countries, the business community and environmental NGOs from the North and the South. A team of policy analysts and experts supported the dialogue. Four two-day workshops were held and there were additional inter-session contacts between the project team and the dialogue participants. Since the emphasis in the global dialogue was on key elements of viable long-term strategies and only one dialogue group could be formed there was no specific sector focus.

However, given its central role as a source of greenhouse gases the energy system received a lot of attention. Special attention in that context was given to the position of countries that are very dependent on fossil fuel resources such as developing countries with large domestic oil and coal reserves. A feature of the global dialogue was the ample attention devoted to the politically sensitive question of equitable distribution between industrialized and developing countries of future efforts to control climate change (Berk et al., 2001).

3.1 Future Image

In terms of global CO_2 emission reductions a situation of 15–25 per cent reduction by 2050 compared to the present was assumed, consistent with a 450 ppm CO_2 stabilization level as indicated above. Energy system characteristics were derived from stabilization studies using the IPCC SRES reference scenarios (Nakicenovic et al., 2000; Morita et al., 2000; Morita et al., 2001; de Vries et al., 2000). In most of the discussions at least two different worlds were assumed: a world with a strong economic market-oriented character and a world with the emphasis on social and environmental values and strong governance (A1 and B1 in the IPCC Emission Scenarios (Nakicenovic et al., 2000)). Two dimensions of such a low CO_2 future were explored in more detail: the energy system (basically the transition to a low-carbon energy system) and global partnerships.

The type of world in which a de-carbonization transition needs to be realized makes a lot of difference. Technological development, resource availability and economic development strongly influence the choice of emission mitigation options. In an A1 world, technological solutions and cost-effectiveness will be strong criteria for selecting options, while in a B1 world social and other environmental aspects will have an important bearing on the selection, resulting in more attention for lifestyle-related options, social impacts and environmental risks of mitigation options. In none of the worlds will a single CO_2 reduction option be sufficient.

As discussed above, some energy options depend more than others on specific conditions that may or may not exist in different future worlds (Morita et al., 2000; UNDP et al., 2000; van Vuuren and de Vries, 2001) and dialogue participants were strongly in favour of them or opposed to them:

- clean coal technologies with CO_2 removal and storage (may not be seen as attractive in societies oriented towards sustainable development unless coupled to hydrogen production);
- solar electricity (may be too expensive in market-oriented worlds);

- nuclear energy (not very likely in sustainable development-oriented worlds);
- large-scale biological carbon sequestration (may be seen as threatening biodiversity and local food supply in sustainable development-oriented worlds unless carefully controlled).

Achieving the kind of global emission reduction we are talking about without a global partnership is impossible. If developing countries are not willing to make a contribution to the global effort in view of their other development priorities, the strict climate change limits cannot be achieved. In that context the question becomes important how addressing climate change can go hand in hand in solving major development challenges (see Table 13.1).

Table 13.1   Characteristics of current and future worlds relevant to realizing global partnerships and social transitions (based on Nakicenovic et al., 2000)

	1990	A1	A2	B1	B2
Poverty eradication	Low	High	Low	Very high	Medium
Reduction of income gaps	Low	High	Low	Very high	Medium
Access to energy	Low	High	Low/ medium	Very high	Medium/ High
Avoiding health effects	Medium	High	Low/ Medium	Very high	High
Global governance	Low/ Medium	High	Low	High	Low
Technology diffusion	Low	Very high	Low	High	Medium
Food security	Medium	High	Low/ medium	High	Medium/ high

3.2   Path Analysis

The development and implementation of the technological options described above will not come about by themselves, at least not within the time frame for stabilizing CO_2 concentrations at 450 ppmv. They require much technological research and development and large capital investments. To realize a transition to a low carbon-intensive society there are also major economic, social and institutional barriers to overcome. At the same time, there are also opportunities, particularly outside the direct realm of climate change, that could be used to help overcome these barriers.

3.3 Barriers to a Decarbonization Transition

There are many barriers to the implementation of options to reduce CO_2 emissions (see also Sathaye et al., 2001; UNDP et al., 2000). A first major barrier, particularly for the private sector, is the uncertainty about climate policies. If investments are to pay off there is a need to know where policies are likely to go. In that respect the recent breakthrough towards ratification of the Kyoto Protocol is very important. This might encourage a pro-active response by the private sector, in particular from those companies that see new market opportunities. The development of new industries may also help to overcome societal resistance to climate change policies.

Second, there are the vested interests of fossil fuel-producing sectors. Although subsidies have been reduced in the past, fossil fuels still receive substantial direct and indirect subsidies. Some of them are related to social concerns, like securing employment or supporting development; others simply reflect the political influence of either the fossil-fuel sectors themselves or energy-intensive industries. The fossil-fuel lobby in the USA against the Kyoto Protocol is a case in point. At the international level it is the OPEC countries that fear the loss of export income.

A third barrier relates to the fear of many industrialized countries of becoming too dependent on energy imports. A shift from coal and oil to natural gas may make many countries even more dependent on imports from a limited number of countries (notably the FSU and countries in the Middle East). This energy dependence makes them strategically vulnerable in times of political tensions or conflict. This is also the reason for subsidizing national energy production and the exploration and development of new, remote (Arctic) and non-conventional fossil reserves (see the Bush administration's Energy Plan). Strategic security considerations are likely to play a role even in a world with a globalized economy (A1), let alone in a more regionalized (A2/B2) type of world.

Fourth, a possibly important barrier, particularly to the development of renewable energy resources and energy saving, is the liberalization of energy markets, which is already taking place in many industrialized countries and would be the rule if the world develops into an A1 type of world. It increases competition between energy suppliers, often resulting into lower energy prices. Intense competition makes it more difficult for production companies to invest in new energy infrastructure (for example for natural gas) and in less commercial or not yet fully commercial energy resources, such as renewables. Moreover, liberalization may also result in a further decline in energy research investments (IEA 1999; UNDP et al., 2000)

Finally, there could be social and environmental barriers to the development of some mitigation options such as large-scale biofuels and bio-sinks, nuclear energy and CO_2 removal and storage. These concerns can be expected especially in environmentally-oriented (B1/B2) types of worlds. Large-scale biomass plantations and bio-sinks may result in monocultures affecting biodiversity and resulting in regional/local competition for land at the expense of food security. Carbon storage will have to overcome concerns about safety risks related to leakage from underground deposits, and ecological risks in the case of deep ocean disposal. Although experts regard hydrogen as no more hazardous than other fuels, public safety concerns may be a barrier to its large-scale use.

3.4   Opportunities for a Decarbonization Transition

There are many opportunities for realizing a transition to a low-carbon energy system. First, the lack of an energy infrastructure and future expansion of energy use in developing countries offers options for 'leap-frogging' to sustainable-energy development paths. Instead of adopting or further expanding conventional fossil fuel-based energy systems, developing countries can avoid sharp increases of CO_2 emissions and other fossil fuel-related environmental problems by opting for socially efficient systems, highly efficient and clean fossil technologies and renewables. Clear examples are the development of public transport systems, the introduction of decentralized combined heat and power systems and renewable energy in off-grid regions.

Second, particularly in developing countries significant reductions in the growth of emissions can be accomplished by pursuing sustainable development policies focusing on reducing other environmental problems, transportation congestion, health and waste problems (biogas) and promoting sustainable land use. Links with other policy goals and co-benefits in designing and promoting climate change strategies are also important in industrialized countries.

Third, while increasing competition, liberalization of energy markets also offers the opportunity to remove fossil fuel subsidies and provide more room for efficient decentralized CHP and renewables. However, this will only work if the environmental costs of fossil fuel use are fully internalized in prices via market mechanisms and/or renewable energy development is adequately promoted via public support of research and development activities and the creation of markets.

Fourth, the development of biofuels offers many countries an important opportunity to reduce their dependence on a limited number of countries for energy imports, particularly of oil. For developing countries it could reduce the burden of oil imports on their balance of payments and may provide additional sources of employment and income from the export of biofuels.

3.5   How to Create a Global Partnership?

It became clear in the dialogue process that developing countries come to this debate with strong views about the current inequities in the world in terms of income, development and use of the global commons. This attitude roughly leads to two different approaches to global partnership:

- an increasing participation approach, which argues that developing countries would be ready to take on responsibilities for greenhouse gas emission reduction if they reached development levels comparable to the industrialized countries;
- an equal rights approach, which argues that the right to emit greenhouse gases should be the same for all citizens of the world, leading to large emission allowances for developing countries even before they actually need them.

In the dialogue the consequences of these approaches were evaluated by the participants with the help of some decision-support tools (the FAIR model; den Elzen et al., 1999 and 2001; Berk and den Elzen, 2001; see also Berk et al., Chapter 9, above), by playing out various scenarios in which developing countries join the global effort. The critical issue in an increasing participation approach (which is more or less the system currently adopted in the Climate Change Convention and the Kyoto Protocol, UNFCCC, 1992 and 1997) is the need to have developing countries join soon if the global reductions for a 450 ppm stabilization of CO_2 are to be achieved. Although different options exist, including the use of relative targets (emissions per unit of GDP) rather than absolute ones (absolute emission levels), accession of developing countries would normally occur at a level of income or level of emission that is much lower than those of industrialized countries when they agreed in 1992 to take action under the UNFCCC. Developing country participants generally see this as unfair unless industrialized countries remain responsible for paying for this action.

Alternatively, an approach that adopts convergence to, for instance, equal per capita emission allowances at some point in the future is generally seen by industrialized country participants as unacceptable because of the perceived high costs or because they are opposed to the principle of equal access to the global commons. Interestingly, it could be shown that a system of convergence of per capita emission allowances over, say, a 20-year time-frame (starting after the Kyoto commitment period, reaching convergence in 2030) would not be extremely expensive for industrialized countries and not unattractive for developing countries (Berk et al., 2001). The reason for the modest costs is that in such a system all countries would receive emission allowances and a global system of emission trading could then be established, allowing huge cost savings to be achieved. For the least developed countries it would generate income for a long time, for other developing countries it would allow them to keep mitigation requirements down if they followed a sustainable development pattern. This points in the direction of approaching the equitable sharing of the global mitigation efforts through the development agenda, an agenda that developing countries rightfully see as their top priority. This would then open up opportunities to design development strategies that promote a transition in the energy system in developing countries towards renewable energy, maybe putting them at the frontier of solar or biomass energy.

3.6 Short-term Action

Short-term action that would move the system in the direction suggested by the strategic observations might involve:

- raising broad public awareness of the climate change problem and acceptance of its (technological) solutions;
- the development of clear and effective global national climate policies, providing incentives to companies and consumers to change their behaviour;
- early ratification of the Kyoto protocol, which now seems realistic;
- an increase and redirection of energy research;
- concentration on some key areas where developed countries could show the way (for example fuel cell car, carbon storage and removal);
- remove obstacles to the effective transfer/diffusion of modern technological knowledge (see also Metz et al., 2000);
- an early start with the Kyoto Mechanisms to establish a price for carbon emissions (now likely after the decisions taken at the 7th Conference of Parties of the UNFCCC);
- assist fossil fuel-dependent countries to restructure their economies and

develop carbon-removal and sequestration technologies (some funding will be made available by the parties to the UNFCCC);
- develop social policies to soften the regional consequences of reduced coal use.

4 ELEMENTS OF A LONG-TERM STRATEGY FOR EUROPE

The European dialogue of the COOL project was organized around two major European sectors, energy production and industrial use, and transport. For each sector between 15 and 20 participants were selected from different backgrounds: private enterprises, utility sectors, consumer organizations, environmental organizations, national and local authorities, research institutes, European institutions, Central and East European authorities and organizations. These participants met four times in one year. Each of the four workshops lasted one to two days and focused on one of the backcasting steps. Only the objective (80 per cent emission reduction by 2050) was decided by the project team and acceptance of that goal was a selection criterion for participation. Every workshop was prepared by the project organization with support from scientific institutions. Between the workshops there was regular contact between the project organization and the participants, particularly for the purpose of getting feedback on drafts, selecting priorities and clarifying preferences (Andersson et al., 2001).

4.1 Energy Production and Industrial Energy Use

Future images
The starting point for the backcasting exercise in the energy group of the European dialogue was the construction of two images of 2050. The first image that was constructed was the so-called *biomass-intensive image*. In this image biomass has the largest share of the 2050 fuel mix (34 per cent). Besides biomass waste, energy crops are expected to require 80 Mha, or 17 per cent of the total surface area in Europe.

The COOL participants expressed a fear that the biomass-intensive image could place excessive demands on the available space in Europe. Large-scale PV plants were suggested as a less land-intensive route for energy production. This resulted in the *solar/hydrogen image*. In this image solar PV accounts for 15 per cent of the 2050 energy mix (compared to 2 per cent in the biomass-intensive image), while only 21 per cent relates to biomass. The solar PV-generated electricity in the solar hydrogen image can also be used to produce hydrogen as a transport fuel, while in the interim hydrogen production from fossil fuel in combination with CO_2 removal and storage can start building a hydrogen infrastructure.

It is noticeable that neither of these two future images foresees a role for coal and nuclear power. But there will be a sharp increase in the use of natural gas and wind power. Both images assume that the economy will continue to grow, while the demand for primary energy remains constant at today's level, indicating a strong improvement in energy efficiency.

It appeared at an early stage of this process that the participants had a strong preference for decentralized production of electricity, fuel and heat, even from an economic (production and distribution costs) and security of supply perspective. Decentralized meant small distributed units operating independently to serve local demand, but also to supply surplus energy to the grid. The operation of each generator may be governed by availability of wind and solar energy and of local fuels as well as by demand for other services than electricity. Still, the operation may be controlled by a centralized control system. This links more strongly to the B types of world in the IPCC scenarios.

Path analysis

To arrive at either of the two images, the growth rates for all the expanding sources (gas, biomass, wind power, solar PV and solar thermal) need to be high. For example, to reach the solar hydrogen image solar PV has to grow by 22 per cent per year. Solar PV can only play a major role in the future if it is forced into the market via regulation and/or subsidies. Major improvements in decentralized renewable energy systems, achieved via a large-scale R&D programme, could reduce the costs even more than anticipated in the learning curves. During the dialogue it was proposed that we may need something like the Manhattan Project or the Apollo Project to bring prices down in some areas of renewable energy technology, in particular solar PV.

One important suggestion made was that carbon removal and storage, if applied, should be based on the 'pre-combustion decarbonization' model. This involves removing the carbon content of fossil fuels to produce hydrogen before their use for energy conversion. Decarbonized ('clean') fossil fuels would then complement renewable energy sources (which are mainly used for electricity in the short term and perhaps for hydrogen towards 2050).

New infrastructure will need to be developed for the distribution of biofuels and hydrogen and for a potential storage system. Major barriers, obstacles and uncertainties were identified in developing and implementing such infrastructure. It is difficult to introduce hydrogen fuel cell cars because no infrastructure is in place to supply the vehicles with the necessary fuels. On the other hand, no hydrogen infrastructure will be developed as long as there is no demand for such alternative fuels. Breaking this seemingly vicious circle requires common visions and co-ordinated action by several actors.

The barriers and obstacles to introducing liquid biofuel systems into existing fuel distribution systems are much lower. Only slightly modified equipment can be used, as has been demonstrated in Brazil's ethanol programme. But institutional co-ordination is a major challenge. Supply and demand have to be introduced simultaneously requiring concerted action by fuel producers, vehicle producers and distributors. Co-operation with the oil industry could be very fruitful for starting experiments with biofuels (limited distribution or mixing fuels).

Short-term actions
Needless to say, it is difficult to imagine the future images being realized unless different actors take early action. Significant uncertainties prevent various actors from investing in innovations that are crucial to take off in the desired direction. Governmental action and co-ordination using existing windows of opportunities is therefore crucial to get out of this impasse. Triggering the hydrogen option is typically a task for the European Commission because the development and implementation of hydrogen infrastructure will take a long time and will require co-operation and co-ordination between many countries and companies in Europe. In the dialogue it was emphasized that EU enlargement is a window of opportunity since this may make both land and labour available to provide bioenergy.

Another short-term window of opportunity is the declining electricity costs in some countries in Europe due to the ongoing liberalization of the electricity markets. This would make it possible to introduce carbon dioxide taxation without industrial producers and consumers being confronted with rising prices. Income from carbon taxation could be used to reduce labour costs, support renewable energy sources and ecologically modernize European industry.

The dialogue has shown that many companies in the private sector are aware that constraints on carbon emissions are likely in the future. Proactive companies have already made efforts beyond the legal obligations. Such voluntary climate protection measures are made easier by transparency and consistency in public policy development as investment decisions can be made with less uncertainty. Such ground-breaking efforts by progressive companies become even more successful and widespread when rewarded by a willingness to pay among final customers and consumers. Not only private customers and consumers are important in this respect; decisive encouragement may come from public bodies acting as customers under public procurement legislation.

4.2   Transport Sector

Future image

Unlike the energy group of COOL Europe, the transport group participants opted for the development of only one future image of 2050 as the basis of further analysis and discussion. The final future image consisted of four dimensions: (1) improved efficiency; (2) fuel substitution; (3) changes in societal structures and patterns; and (4) changes in awareness, values and lifestyles. Improved energy efficiency and new fuels directly affect emissions (emissions per unit of transport), while new patterns of human activities and values and lifestyles mainly have an impact on transport volume.

In the future image for 2050 the transport system is characterized by a large variety of niche vehicles (for example small electric city vehicles), all-purpose cars and new systems such as personal electric vehicles that can link to each other and form trains that move along special tracks. There is no single transport mode that dominates the system to the same extent as the private all-round car did at the beginning of the twenty-first century. Another prominent feature is the spread of inter-modal transport with smooth and short transitions between modes. IT is widely used in intelligent traffic control and information systems, and also for flexible road pricing. The energy efficiency in the transport sector is high.

The future image served its purpose very well: it was a starting point for discussion, to get an idea of the 'agenda' for transport in 2050. However, the group was very much aware that this image was only one of many possible futures.

Path analysis

Subsequently, in the path analysis a range of options was developed for each of the four dimensions identified, providing a path from the future image to the present. For each of the options, barriers, opportunities, uncertainties, an indicative time-path, advocacy coalitions, and necessary conditions were identified:

- Many options designed to improve efficiency can be taken in the short term. They are relatively easy to develop and implement and do not involve many uncertainties. The appropriate actors can also be readily identified.
- Fuel substitution involves more difficult choices, greater uncertainties

and more complex coalitions of actors. It is of greater significance in the medium term. However, it also offers more scope for radical change.

- Changes in structures and patterns require a long period to take effect and involve substantial uncertainties. The changes also depend to a large extent on appropriate actions by a wide range of actors in other sectors and need supportive conditions from the external environment.
- Changes in awareness, values and lifestyles are, on the one hand, difficult to influence since they are dependent on a wide range of factors and complex interactions within society. On the other hand, lifestyles and value-patterns change constantly, not always in a favourable direction from an environmental point of view. Such changes contain a huge potential to contribute to radical changes in CO_2 emissions.

Short-term actions

In order to cut CO_2 emissions by 80 per cent by 2050 many of the technological innovations, changes in lifestyles, structural transformations and institutional and policy reforms will have to start in the short term. A distinction was made between actions by the (private) sector itself and policy actions. Actions in the private sector included improved co-operation between transport producers/providers and customers, using environmental aspects as a marketing tool, shared use of environmental and logistical information systems, investment in new services and multimodal companies and including transport in environmental reporting.

It was noted that energy is currently too cheap for a move in the direction of climate neutral transport. One of the tasks identified for policymakers was to initiate the debate within the sector on the internalization of external costs. Measures like the kilometre tax, road pricing and an emission trading system were seen as particularly important in this respect. With regard to R&D, European policy should play a role in providing room for demonstration projects and stimulating co-operation between cities to increase the market for new transport technologies.

4.3   European Institutional Conditions

A number of European institutional conditions for facilitating these transitions in the energy and transport sectors were identified. First, it is not feasible for the EU to move towards the 80 per cent target unilaterally. Therefore, the EU could elaborate a vision about how the climate policy regime should develop in the long term to meet the ultimate objectives of the UNFCCC and of the roles of Europe, of the different European sectors such as energy and transport and of the EU institutions.

Second, with these gigantic challenges ahead it is desirable to both strengthen the European Commission's capacity to deal with long-term climate policy, as well as to strengthen the co-operation and collaboration with the private sector and other stakeholders. It is, for instance, important to establish new institutions with responsibility for energy efficiency in the liberalized energy markets in Europe.

Finally, the enlargement of the EU brings new challenges (for instance in terms of the inefficient energy use of various systems and increasing regional transport), new opportunities (for instance in terms of biomass production in the accession countries in Eastern Europe and bullet trains instead of continental flights) and new institutional requirements (for instance in terms of majority voting for major tax reforms).

5   LESSONS LEARNED

In terms of important elements of a long-term strategy the COOL project was particularly useful in identifying key barriers and opportunities for achieving low carbon futures and some important short term actions. Some of the most important finding were:

- *Uncertainties in short- and long-term climate policy*: the existing uncertainty regarding the implementation of the Kyoto Protocol (the recent breakthrough gives hope) and lack of clarity concerning the level at which greenhouse gas concentrations are going to be stabilized, even on a provisional basis, form major obstacles for the private sector to direct their investments in low-carbon energy and transportation technologies. Ratifying the Kyoto Protocol (now likely), establishing a price for carbon emissions through the Kyoto mechanisms and some unilateral (EU) objectives for further emission reductions in the 2030 time-frame would be important short-term steps.
- *Confusing signals in energy pricing and importance of low-carbon energy innovation*: not including external costs of environmental damage, the liberalization of energy markets, declining R&D funding, lack of initiatives for building a different energy infrastructure (for example hydrogen networks) and existing coal subsidies are contributing to a general lack of incentives for introducing low-carbon energy systems. Creating price signals through ecological tax reform or tradable emission permits and active involvement of governments in funding and co-ordinating innovation is required.
- *Vested interests and energy security concerns*: the necessary rapid shift

from traditional fossil fuels to natural gas, renewable energy and hydrogen systems is hampered by fossil fuel interests (OPEC, major oil companies, coal mining). Wind, biomass and solar energy can contribute to diversification and play a positive role in energy security concerns. Carbon removal and storage technologies can help to build 'clean fossil' systems as a bridge to a future renewable energy society. Political decisions however have to be taken.

- *Lack of awareness, public resistance and lifestyle preferences*: a complex array of factors contributes to a much slower uptake of technological and social innovations than would be technically and economically possible. Although there are no simple solutions that would rapidly overcome these obstacles, a broad set of information and public participation activities combined with social science research can certainly help to deal with these problems.

- *International governance and general socio-economic policy*: rapid diffusion of new energy technologies around the world, agreement on international co-operation in dealing with climate change and providing the necessary conditions for countries to shift their development patterns in a sustainable direction call for many deliberate steps in national and international policymaking. Strengthening and co-ordinating inter- national institutions and bilateral and regional co-operative programmes is a necessary step.

In addition, conclusions can be drawn from the COOL project regarding the scientific methodology applied, especially the backcasting approach and the additional value of a participatory approach in integrated and policy assessments. The backcasting methodology proved a valuable tool in reaching the final substantial results identified above. However, two qualifications should be made:

- In applying a backcasting methodology the path analysis backwards proved especially difficult. While this analysis provided valuable information on barriers, opportunities and crucial timeframes for developing specific technologies, this proved far more difficult for behavioural and institutional changes. In addition, a path analysis over the full range of social, institutional, economic and technological transformations from 2050 back to the present would require enormous resources. This clearly shows the limitations of backcasting approaches.

- In an overall backcasting exercise forecasting and scenario studies were indispensable on specific issues and for shorter timeframes. Both approaches seem to be, and should be, used complementarily rather than competitively.

With respect to the participatory approach some clear conclusions can be drawn. It is not so much that participation by various stakeholders provided many new, unknown or surprising elements of a long-term strategy for European and global climate policy. The result of the participatory process within the COOL project proved to be:

- the identification and clarification of the various (technical and non-technical) barriers to radical and long-term greenhouse gas emission reduction strategies;
- the building of support for, and to a large extent even consensus among, participants for specific options, strategies and necessary short-term actions;
- an element of learning among participants: first, on the diversity of possible options to combat climate change, on the scientific uncertainties and on the technical, economic, institutional and social barriers to implementing these options. Second, on the logic and rationality of positions and interests of other participants, which is crucial for identifying feasible strategies.

REFERENCES

Andersson, M., W. Tuinstra and A.P.J. Mol (eds) (2001), *Climate Options for the Long Term-European Dialogue Final Report (Final report COOL part C)*, Wageningen: Wageningen University, Department of Social Sciences.

Berk, M.M and M.G.J. den Elzen (2001), 'Options for Differentiation of Future Commitments in Climate Change Mitigation: How Can we Enhance Participation in Global Emission Control?', *Climate Policy*, 1(4): 465-480.

Berk, M., J. van Minnen, B. Metz and W. Moomaw (2001), *COOL Global Dialogue – Synthesis Report (volume D)*, Bilthoven: Netherlands National Research Program on Global Air Pollution and Climate Change.

De Vries, H.J.M., J. Bollen, M.G.J. den Elzen, M.A. Janssen, G.J. Kreileman and R. Leemans (2000), 'Greenhouse Gas Emissions in an Equity-, Environment- and Service-Oriented World: An IMAGE-Based Scenario for the Next Century', *Technological Forecasting and Social Change*, **63**, 137–74.

Den Elzen, M., M.M. Berk, M. Schaeffer, O.J. Olivier, C. Hendriks and B. Metz (1999), *The Brazilian Proposal and Other Options for International Burden Sharing: An Evaluation of Methodological and Policy Aspects using the FAIR Model*, Report No. 728001011, Bilthoven: Netherlands National Institute of Public Health and the Environment (RIVM).

Den Elzen, M., M.M. Berk, S. Both, A. Faber and R. Oostenrijk (2001), *FAIR 1.0: An Interactive Model to Explore Options for Differentiation of Future Commitments in International Climate Policy Making*, Report No. 728001012, Bilthoven: Netherlands National Institute of Public Health and the Environment (RIVM).

Douglas, M. and A. Wildavsky (1983), *Risk and Culture. An Essay on the Selection of Technological and Environmental Dangers*, Berkeley: University of California Press.

Drehborg, K. (1996), 'Essence of backcasting', *Futures*, **28**(9), 813–28.

Dryzek, J.S. (1987), *Rational Ecology. Environment and Political Economy*, Oxford: Basil Blackwell.

Dunn, W.N. (1994), *Public Policy Analysis*, 2nd edn, Englewood Cliffs: Prentice Hall.

Fisher, F. (1990), *Technocracy and the Politics of Expertise*, Newbury Park: Sage.

Funtowics, S.O. and J.R. Ravets (1993), 'Science for the Post-normal Age', *Futures*, **25**(7), 739–55.

Hisschemöller, M. and R. Hoppe (1996), 'Coping with Intractable Controversies: The Case for Problem Structuring in Policy Design and Analysis', *Knowledge and Policy: The International Journal of Knowledge Transfer and Utilization*, **8**(4), 40–61.

Hisschemoller, M. and M. van de Kerkhof (eds) (2001), *Climate OptiOns for the Long Term – Nationale Dialoog* (Final Report COOL, Part B), Amsterdam: Institute for Environmental Studies.

International Energy Agency (1999), *IEA Energy Technology R&D Statistics: 1974–1997*, Paris.

IPCC (1995), *IPCC Second Assessment Synthesis of Scientific–Technical Information Relevant to Interpreting Article 2 of the UNFCCC*, Geneva. http://www.ipcc.ch.

Irwin, A. (1995), *Citizen Science. A Study of People, Expertise and Sustainable Development*, London: Routledge.

Mayer, I. (1997), *Debating Technologies. A Methodological Contribution to the Design and Evaluation of Participatory Policy Analysis*, Tilburg: Tilburg University Press.

Metz, B., O. Davidson, J.W. Martens, S. van Rooyen and L. van Wie (eds) (2000), *IPCC Special Report on the Methodological and Technological Aspects of Technology Transfer*, Cambridge: Cambridge University Press.

Mol, A.P.J. and G. Spaargaren (2000), 'Ecological Modernisation Theory in Debate: A Review', *Environmental Politics*, **9**(1), 17–49.

Morita, T., N. Nakicenovic and J. Robinson (2000), 'Overview of Mitigation Scenarios for Global Climate Stabilizing Based on New IPCC Emission Scenarios (SRES)', *Environmental Economics and Policy Studies*, 3(2).

Morita, T., J. Robinson, A. Adegbulugbe, J. Alcamo, D. Herbert, E. Lebre la Rovere, N. Nakicenivics, H. Pitcher, P. Raskin, K. Riahi, A. Sankovski, V. Solkolov, B. de Vries and D. Zhou (2001), 'Greenhouse Gas Emission Mitigation Scenarios and Implications', in B. Metz, O. Davidson, R. Swart and J. Pan (eds), *Climate Change 2001: Mitigation; Contribution of Working Group III to the Third Assessment Report of the IPCC*, Cambridge: Cambridge University Press.

Nakicenovic et al. (2000), *IPCC Special Report on Emissions Scenarios*, Cambridge: Cambridge University Press.

Rotmans, J. et al. (2000), *Transities en transitiemanagement. De casus van een emissiearme energievoorziening*, Maastricht: International Centre for Integrative Studies.

Sathaye, J., D. Bouille, D. Biswas, P. Crabbe, L. Geng, D. Hall, H. Imura, A. Jaffe, L. Michaelis, G. Peszko, A. Verbruggen, E. Worrell and F. Yamba (2001), 'Barriers, Opportunities and Market Potential of Technologies and Practices', in B. Metz, O. Davidson, R. Swart and J. Pan (eds), *Climate Change 2001: Mitigation; Contribution of Working Group III to the Third Assessment Report of the IPCC*, Cambridge: Cambridge University Press.

Schot, J. (1992), 'Constructive Technology Assessment and Technology Dynamics: The Case of Clean Technologies', Science, Technology and Human Values, 17(1), 36–56.

Seley, J.F. (1983), *The Politics of Public Facility Planning*, Lexington: Lexington Books.

UNDP, UNDESA, WEC (2000), *World Energy Assessment*, New York: UNDP.

UNFCCC (1992), *The United Nations Framework Convention on Climate Change*, http://www.unfccc.int/resources.

UNFCCC (1997), *Kyoto Protocol to the United Nations Framework Convention on Climate Change*, http://www.unfccc.int/resources.

Van Thijn, E. (1997), *Politiek en bureaucratie: baas boven baas*, Amsterdam: Van Gennep.

Van Vuuren, D.P. and De Vries, H.J.M., (2001), 'Mitigation Scenarios in a World Oriented at Sustainable Development: The Role of Technology Efficiency and Timing', *Climate Policy*, 1, 189–10.

Wynne, B. (1996), 'May the Sheep Safely Graze? A Reflexive View of the Expert–Lay Knowledge Divide', in S. Lash et al. (eds), *Risk, Environment and Modernity. Towards a New Ecology*, London: Sage, pp. 44–83.

Yohe, G., (2001), 'Mitigative Capacity: The Mirror Image of Adaptive Capacity on the Emissions Side', *Climatic Change*, **46**, 371–90.

Index